Math Mammoth
Grade 7 Review Workbook

By Maria Miller

Contents

Introduction

Math Mammoth Grade 7 Review Workbook is intended to give students a thorough review of seventh grade math, following the main areas of Common Core Standards for grade 7 mathematics and typical pre-algebra study topics. The book has both topical as well as mixed (spiral) review worksheets, and includes both topical tests and a comprehensive end-of-the-year test. The tests can also be used as review worksheets, instead of tests.

You can use this workbook for various purposes: for summer math practice, to keep the child from forgetting math skills during other break times, to prepare students who are going into eighth grade or algebra 1, or to give seventh grade students extra practice during the school year.

The topics reviewed in this workbook are:

- algebra
- integers
- one-step equations
- rational numbers
- equations and inequalities
- ratios and proportions
- percent
- geometry
- Pythagorean Theorem
- probability
- statistics

In addition to the topical reviews and tests, the workbook also contains many cumulative (spiral) review pages.

The content for these is taken from the *Math Mammoth Grade 7 Complete Curriculum*. However, the content follows a typical pre-algebra course, so this workbook can be used no matter which math curriculum you follow.

Please note this book does not contain lessons or instruction for the topics. It is not intended for initial teaching. It also will not work if the student needs to completely re-study these topics (the student has not learned the topics at all). For that purpose, please consider the *Math Mammoth Grade 7 Complete Curriculum*, which has all the necessary instruction and lessons.

I wish you success with teaching math!

Maria Miller, the author

The Language of Algebra Review

1. Find the value of these expressions.

a. $(6+4)^2 \cdot (12-9)^3$	**b.** $3 \cdot (5 - (7-5))$	**c.** $\dfrac{(5-3) \cdot 2}{2^3} + 7$

2. Name the property of arithmetic illustrated by the fact that $(5 \cdot z) \cdot 3$ is equal to $5 \cdot (z \cdot 3)$.

3. Evaluate the expressions.

a. $100 - 2x^2$, when $x = 5$	**b.** $\dfrac{2s}{s^3 + 3}$, when $s = 4$

4. Which equation matches the situation? *Hint: give the variable(s) some value(s) to test the situation.*

a. The shorter beam (length l_1) is 1.5 meters shorter than the longer beam (length l_2).

$l_1 = 1.5 - l_2$	$l_2 = 1.5 - l_1$	$l_2 = l_1 - 1.5$	$l_1 = l_2 - 1.5$

b. The dog lost 1/6 of its original weight (w), and weighs now 23 kg.

$\dfrac{w}{6} = 23$	$\dfrac{5w}{6} = 23$	$\dfrac{6w}{5} = 23$	$w - 1/6 = 23$	$w - 5/6 = 23$

5. Is subtraction commutative? In other words, is it true that $a - b$ has the same value as $b - a$, no matter what values we use for a and b? Explain your reasoning.

6. Write a SINGLE expression to match these situations.

a. A pair of jeans costs p dollars. The jeans are now discounted by 1/5 of that price.
Write an expression for the discounted price.

b. It costs Mandy $0.18 to drive her car one mile.
How much does it cost her to drive y miles? Write an expression.

c. The pet store sells dog food in bags of two different sizes: 3-kg and 8-kg.
What is the total weight of n of the smaller bags and m of the larger bags?

7. Simplify the expressions.

a. $x + 2 + x + x$	b. $x \cdot 2 \cdot x \cdot x \cdot x$	c. $8v + 12v$
d. $8v \cdot 12v$	e. $4z \cdot 9z \cdot z$	f. $f + 2f + 10g - f - 4g$

8. **a.** Sketch a rectangle that is $5x$ tall and $2x$ wide.

   b. What is its area?

   c. What is its perimeter?

9. Use the distributive property to multiply.

a. $12(v - 9)$	b. $3(a + b + 2)$	c. $3(0.5t - x)$

10. Draw a diagram of two rectangles to illustrate that the product $11(x + 7)$ is equal to $11x + 77$.

11. Fill in the table.

Expression	the terms in it	coefficient(s)	Constants
a^8			
$2x + 9y$			

12. The perimeter of a regular pentagon is $30s + 45$. How long is one side?

13. Factor these sums (write them as products). Think of divisibility!

a. $48x + 12 =$	b. $40x - 25 =$
c. $6y - 2z =$	d. $56t - 16s + 8 =$

The Language of Algebra Test

1. Write an expression with three terms. The coefficient of the first term is 2 and of the second term is 5. The last term is the constant 9. The variable part of the first term is s squared, and the variable of the second term, t.

2. Evaluate the expressions.

a. $2(7 - x)^2$, when $x = 2$	**b.** $\dfrac{1}{g} + \dfrac{g + 1}{3}$, when $g = 6$

3. Name the property of arithmetic illustrated by the fact that $5(z + 4)$ is equal to $5z + 20$.

4. Draw a diagram of two rectangles to illustrate that the product $5(z + 4)$ is equal to $5z + 20$.

5. Write each expression as a product (factor it).

a. $7x + 21 = \underline{\quad}(x + \underline{\quad})$	**b.** $24k + 80 =$

6. Simplify the expressions.

a. $v + 5 + v + v + v$	**b.** $v \cdot 5 \cdot v \cdot v \cdot v$	**c.** $8x + 5 - 3x - 2$

7. Write an equation and solve it using guess and check.

    a. Seven times the quantity x minus one equals 14.

    b. Two less than x squared equals 23.

8. Write an expression for each situation.

    a. Abigail bought x bags of nuts for $3 a bag. She paid with a $50 bill. What was her change?

    b. A pair of jeans that costs p dollars is discounted by 1/10 of its price. What is the discounted price?

9. **a.** Write and simplify an expression for the total area of the shape.

    b. Evaluate your expression when $x = 2$ cm.

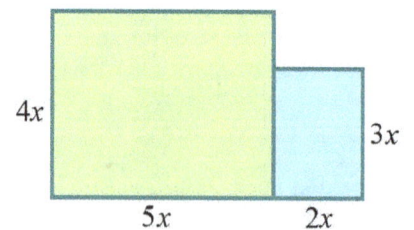

4x   3x

5x   2x

    c. Write an expression for the perimeter of the shape.

Integers Review

1. Match the equations with the situations and complete the missing parts.

   a. A ball was dropped from 18 ft above sea level; it fell 12 ft.
   Now the ball is at _____ ft.

   b. John had a $12 debt. He earned $18. Now he has _____.

   c. John had $12. He had to pay his dad $18. Now he has _____.

   d. A diver was at the depth of 18 ft. Then he rose 12 ft.
   Now he is at _____ ft.

   e. The temperature was −12°C and fell 18°. Now it is _____ °C.

   $12 − 18 = $ _____

   $−12 + 18 = $ _____

   $−18 + 12 = $ _____

   $−12 − 18 = $ _____

   $18 − 12 = $ _____

2. Compare the integers using $>$ or $<$. Include the units, too.

   a. The temperature inside the fridge is 5°C.
   In the freezer, it is 12°C below zero.

   b. Andy has a debt of $400. Harry owes $250.

3. Add or subtract.

a. $(−12) + (−1) = $ _____	**b.** $−12 − (−1) = $ _____	**c.** $7 − 12 = $ _____

4. Complete the equations, using <u>one positive</u> and <u>one negative</u> integer. There are many possible solutions.

a. _____ + _____ = −2 _____ + _____ = −2	**b.** _____ + _____ = 0 _____ + _____ = 0	**c.** _____ + _____ = 3 _____ + _____ = 3

5. Interpret the absolute value in each situation.

   a. A shark is swimming at the depth of 14 m. $|−14 \text{ m}| = $ _____ m

   Here, the absolute value shows _____

   b. Shelley's balance is −$31. $|−\$31| = \$$ _____

   Here, the absolute value shows _____

6. Write using symbols, and simplify if possible.

   a. the opposite of 8

   b. the opposite of −100

   c. the opposite of the sum 2 + 5

   d. the absolute value of negative 45

7. Write an addition or subtraction sentence to match the temperature change.

   a. The temperature was −2°C. Then it rose 4 degrees. Now it is _____.

   b. The temperature was −11°C. Then it rose 5 degrees. Now it is _____.

   c. The temperature was 2°C. It dropped 8 degrees. Now it is _____.

   d. The temperature was −3°C. It dropped 8 degrees. Now it is _____.

$^-2 + 4 =$ _____

8. Plot these inequalities on the number line.

← ‖ + + + + + + + + + + + + + → −8 −7 −6 −5 −4 −3 −2 −1 0 1 2 3 4	← ‖ + + + + + + + + + + + + + → −8 −7 −6 −5 −4 −3 −2 −1 0 1 2 3 4
a. $x \geq -2$	**b.** $x < 2$

9. Add.

a. $(-2) + 7 + (-7) + (-1) =$ _____	**b.** $4 + (-10) + (-12) + 1 =$ _____

10. Iodide is an ion with 53 protons and 54 electrons.
    What is the total electric charge of this ion?

11. The top of a fishing net is at a depth of 6 m below the surface of a lake, and later it is lowered to the depth of 8 m. Write an expression for the distance between these two depths using negative integers.

12. Solve.

a. $21 + (-48) =$	**b.** $41 + (-38) =$	**c.** $-610 + 900 =$

13. Change each subtraction into an addition and solve.

a. $1 - (-7)$	**b.** $2 - (-11)$	**c.** $-20 - (-6)$	**d.** $3 - 8$
↓	↓	↓	↓
____ + ____ = _____	____ + ____ = _____	____ + ____ = _____	____ + ____ = _____

14. Which king ruled the Persian Empire longer, Xerxes I, who ruled from 486 to 465 BC, or Darius II, who ruled from 424 to 404 BC?

15. Multiply.

a. $-2 \cdot (-4) =$ _____	**b.** $(-3) \cdot (-8) =$ _____	**c.** $(-3) \cdot 3 \cdot (-1) =$ _____
$-2 \cdot 4 =$ _____	$7 \cdot (-12) =$ _____	$-7 \cdot (-2) \cdot (-2) =$ _____

16. True or false?

    a. Any integer more than 6 has an absolute value more than 6.

    b. Any integer less than 6 has an absolute value less than 6.

    c. A number and its opposite have the same absolute value.

    d. The absolute value of the opposite of a number is the same as
        the opposite of the absolute value of the same number.

17. Divide.

a. $-10 \div (-5) =$ _____ $24 \div (-3) =$ _____	**b.** $(-12) \div (-4) =$ _____ $21 \div (-3) =$ _____	**c.** $-56 \div 7 =$ _____ $-120 \div (-10) =$ _____

18. Find the missing numbers.

a. $-5 \cdot$ _____ $= -30$	**b.** $2 \cdot$ _____ $= -18$	**c.** $-8 \cdot$ _____ $= 48$
d. $-42 \div$ _____ $= 6$	**e.** $-64 \div$ _____ $= -8$	**f.** $81 \div$ _____ $= -9$

19. Solve the equations by thinking of multiplication tables.

a. $5y = -100$ $y =$ _____	**b.** $-4b = -48$ $b =$ _____	**c.** $\dfrac{35}{-5} = y$ $y =$ _____

20. Give a real-life situation for the product $3 \cdot (-10)$.

21. Divide and simplify if possible.

a. $1 \div (-6)$	**b.** $-3 \div 15$	**c.** $-6 \div (-7)$

22. Find the value of the expressions when $x = -3$ and $y = 4$.

a. x^2	**b.** $-5xy$	**c.** $2 - (y + x)$

Integers Test

1. Illustrate the sum $-5 + 7$ in two ways: (1) using counters, and (2) using a number line.

2. Write a sum where you add a number and its opposite.

3. Add or subtract.

a. $11 + (-8) =$	b. $11 - (-8) =$	c. $-11 - 8 =$	d. $-11 + 8 =$

4. Add or subtract.

a. $61 + (-82) =$	b. $-55 - (-29) =$	c. $43 - 189 =$	d. $72 + (-99) =$

5. Explain a real-life situation for the calculation $-80 \text{ m} - 30 \text{ m} = -110 \text{ m}$.

6. Write an addition or subtraction, and solve.

Aiden owed $21 on his credit card. Then he paid $15. Then he made a purchase for $35. Then he made another payment of $50. What is his balance now?

7. Simplify.

a. $|-14| =$	b. $-|-4| =$	c. $-(-9) =$	d. $-(13 - 8) =$

8. Allison's mom designed a point system for Allison where she would get positives for doing her chores and school work well and negatives for doing her chores and school work poorly.

   On Tuesday, Allison "earned" five negative points. On Monday, her final tally had been 6 points. How many points better did Allison do on Monday than on Tuesday?

9. **a.** Write an expression for the distance between -2 and -18.

   b. Write an expression for the distance between x and 5.

   c. Evaluate the expression from (b) when $x = -2$.

10. Fill in the missing numbers.

a. $-7 \cdot$ _____ $= -35$	b. $2 \cdot$ _____ $= -100$	c. $-50 \div$ _____ $= -10$

11. Divide and simplify to the lowest terms.

a. $15 \div (-3) =$	b. $-2 \div (-10) =$
c. $-20 \div 6 =$	d. $72 \div (-48) =$

12. Find the value of the expressions when $x = -3$ and $y = 4$.

a. $xy - 2$	b. $x^2 + 1$	c. $-2(y - 5)$

13. Solve the equations by thinking logically.

a. $-8y = 96$	b. $-20a = -300$	c. $36 \div w = -6$
$y =$ _____	$a =$ _____	$w =$ _____

15

Mixed Review 1

1. Find the value of the expressions.

    a. $(3 \text{ cm})^3$ **b.** $(1.5 \text{ in})^2$ **c.** $(10 \text{ m})^3$

2. Write an expression for the volume of a cube with edge $2s$.

3. Evaluate the expression from exercise 2 when $s = 1.7$ cm.

4. Write an equation. Then solve it using mental math.

    a. 201 decreased by a number is 167.

    b. The product of a number and 7 equals 7/24.

Equation	Solution

5. Is subtraction commutative? Explain your reasoning using an example.

6. Name the property of arithmetic illustrated by the equation $(n + 2) + 5 = n + (2 + 5)$.

7. Fill in the table.

Expression	Terms in it	Coefficient(s)	Constant(s)
$(5/6)s^2$			
$x + 2y + 8$			
$p \cdot 46$			

8. Write an expression for the area in two ways, thinking of the overall rectangle or its component rectangles.

____ (____ + ____) and

3 · ____ + ____ · ____

9. Use the distributive property "backwards" to write the expression as a product (it is called factoring).

a. $3x + 6 =$ ____ (____ + ____)

b. $10z - 20 =$ ____ (____ − ____)

10. Write using symbols, and simplify if possible.

    a. The opposite of the absolute value of 2

    b. The absolute value of the opposite of 2

11. Write an inequality. Use negative integers where appropriate.

    a. The ditch is at least 2 ft deep.

    b. Ashley owes me no more than $100.

    c. The freezer temperature shouldn't be colder than 20 Celsius degrees below zero.

12. The fence that surrounds Emily's yard is x feet long. Emily has painted 1/4 of it. Write an expression for the length of the fence that is *not* painted.

13. Sketch a rectangle with an area of $3x + 21$.

14. Simplify the expressions.

a. $t + 2t + 9 - t$	**b.** $x \cdot 4 \cdot x \cdot x$	**c.** $5 \cdot y \cdot y \cdot x \cdot 2$

Mixed Review 2

1. Rewrite each expression using a fraction line, then simplify.

a. $7 \div 8 \cdot 4$	**b.** $5 \cdot 2 \div 10 + 1$	**c.** $(10 + 3) \div (8 - 1)$

2. Evaluate the expressions. (Give your answer as a fraction or mixed number, not as a decimal.)

a. $\dfrac{x+2}{x-2}$, when $x = 21$	**b.** $3s^2 - 2t^2$, when $s = 10$ and $t = 3$

3. Name the property of arithmetic illustrated by the equation $(5x)y = 5(xy)$.

4. There are two broomsticks, one wooden and one metal.

   a. Choose two variables to denote the lengths of the two broomsticks.

   Let _____ be the length of the wooden broomstick.

   Let _____ be the length of the metal one.

   b. Write an equation that matches the sentence "The wooden broomstick is 20 cm longer than the metal one."

5. **a.** Circle the equation that matches the situation.

   Let p be the normal price of one sun hat in a clothing store.
   The store owner decides to discount them by $5 each.
   A customer buys three sun hats. The total cost is $16.80.

$3(p - \$5) = \$16.80$	$p - \$5 = 3 \cdot \$16.80$
$3p - \$5 = \$16.80$	$3(p - 0.5) = \$16.80$

   b. How much would one sun hat have cost before the discount?
   Solve this problem using any strategy. You don't have
   to use the equation.

6. Simplify the expressions.

a. $6p + 2 + 5p - 1$	b. $6p \cdot p \cdot 7$	c. $f \cdot 2f \cdot 2f \cdot f \cdot 3$

7. Simplify.

a. $|-71|$	b. $-|-2|$	c. $|-9 + 5|$	d. $-(-84)$
e. $|-9| + |-5|$		f. $|-9| - |5|$	

8. Write an inequality. Use negative integers where appropriate.

a. This hill is at least 200 ft high.

b. Liz owes more than $120.

c. A maximum of 8 items per customer.

d. The ride is only for children that are up to 120 cm tall.

9. Find the missing numbers. You can think of jumps on the number line.

a. $5 - \rule{2cm}{0.4pt} = {}^-2$	c. $2 + \rule{2cm}{0.4pt} = {}^-4$	e. ${}^-30 + \rule{2cm}{0.4pt} = {}^-40$	g. ${}^-51 + \rule{2cm}{0.4pt} = 0$
b. ${}^-1 - \rule{2cm}{0.4pt} = {}^-19$	d. $4 - \rule{2cm}{0.4pt} = 5$	f. $0 - \rule{2cm}{0.4pt} = {}^-49$	h. ${}^-9 + \rule{2cm}{0.4pt} = {}^-7$

10. Answer the questions about the pattern.

Step   1   2   3   4   5

a. Draw steps 4 and 5.

b. How do you see this pattern growing?

c. How many flowers will there be in step 39?

d. What about in step n?

Solving One-Step Equations Review

1. Solve. Check your solutions.

a. $\quad x + 7 \;\; = \;\; -6$	**b.** $\quad -x \;\; = \;\; 5 - 9$
c. $\quad 2 - x \;\; = \;\; -8$	**d.** $\quad 2 - 6 \;\; = \;\; -z + 5$
e. $\quad \dfrac{x}{11} \;\; = \;\; -12$	**f.** $\quad \dfrac{q}{-3} \;\; = \;\; -40$
g. $\quad 100 \;\; = \;\; \dfrac{c}{-10}$	**h.** $\quad \dfrac{a}{5} \;\; = \;\; -10 + (-11)$

2. *Write an equation for the problem. Then solve it.*

Alex bought three identical solar panels and paid a total of $837.
How much did one cost?

Equation:

3. *Write an equation for the problem. Then solve it.*

   Andrew pays 1/7 of his salary in taxes. If he paid $187 in taxes, how much was his salary?

   Equation:

4. Use the formula $d = vt$ to solve the problem.

If you can bicycle at a speed of 20 km/h, how long will it take you to bicycle from the shopping center to a dentist's office, a distance of 1.2 km?	d = v t ↓ ↓ ↓

5. Taking a bus, Emily can get to the community center that is 1.5 km from her home in 3 minutes. What is the average speed of the bus, in kilometers per hour?

6. Ed skates on his skateboard to school, which is 2 miles away. He travels half of the distance at a speed of 12 mph and the rest at a speed of 15 mph. How long does it take him to get to school?

Solving One-Step Equations Test

1. Solve. Check your solutions.

a. $x + 8 = -13$		**b.** $4 - (-2) = -y$	
c. $18 - x = -1$		**d.** $2 - 6 = -z + 5$	
e. $\dfrac{x}{10} = -17 + 5$		**f.** $-13 = \dfrac{c}{-7}$	

Write an equation for each problem. Then solve it. Don't write just the answer.

2. **a.** Seven pounds of chicken costs $32.41. How much does one pound cost?

b. Noah's suitcase is 4.6 kg heavier than Bill's. If Noah's suitcase weighs 28.7 kg, then how much does Bill's weigh?

3. Use the formula $d = vt$ to solve the problem.

A ferry travels at a constant speed of 18 km/h.
How long will it take to cross a river, a distance of 600 *meters*?

$$d = v \quad t$$
$$\downarrow \quad \downarrow \quad \downarrow$$

4. How far can you travel in 1 hour 25 minutes, bicycling at a constant speed of 15 km/h?

5. **a.** Find the average speed of an airplane that flies 2900 miles in 4 1/2 hours.

   b. Find the average speed of the same airplane in kilometers per hour.
      Use the conversion 1 mi = 1.609 km.

Mixed Review 3

1. Write an expression.

   a. 10 less than x squared.

   b. The quotient of 154 and k cubed.

   c. The quantity x plus 2 to the fifth power.

   d. x plus 2 to the fifth power.

2. The sides of a square are $(x + 2)$ long.

   a. Sketch the square.

   b. Write an expression for the area of the square.

   c. Write an expression for the perimeter of the square.

   d. Evaluate your expression for the area of the square when $x = 1.5$.

3. Draw a number line jump for each addition or subtraction.

   a. $-2 + 6 =$ _____   **b.** $-3 - 5 =$ _____

4. Draw counters for the addition $3 + (-5)$. Explain how to perform the addition using the counters.

5. Solve.

a. $89 + (-35) =$	**b.** $-45 + (-29) =$	**c.** $-78 + 60 =$

6. Change each addition into a subtraction or vice versa. Then solve whichever is easier. Sometimes changing the problem will not make solving it easier, but the aim of this exercise is to practice making the change.

a. $-2 + (-18)$	**b.** $56 - (-34)$	**c.** $-14 + (-24)$	**d.** $2 + 9$
↓	↓	↓	↓
___ − ___ = ___	___ + ___ = ___	___ − ___ = ___	___ − ___ = ___

7. Write comparisons using $>$, $<$, and integers. Include the units, too.

   a. The temperature at the North Pole is -34 degrees Celsius, whereas in New York, it is -8 degrees Celsius.

   b. The total electric charge of 12 electrons is $-12e$.
   The total electric charge of 3 protons is $+3e$.

8. Name the property of arithmetic illustrated by the equation $2x = x \cdot 2$.

9. Evaluate the expression $|a - b|$ for the given values of a and b. Check that the answer you get is the same as if you had used a number line to figure out the distance between the two numbers.

a. a is 8 and b is 54	**b.** a is -12 and b is -5

10. Describe a situation where one person has a positive account balance and another has a negative balance, and the one person's balance is $30 more

11. Use the distributive property "backwards" to write the expression as a product.

   a. $42s + 28 = $ _____ (_____ + _____)   **b.** $54z - 18 = $ _____ (_____ − _____)

12. Solve the equations.

a. $\dfrac{x}{-5} = 35$ $x = $ _____	**b.** $\dfrac{35}{y} = -5$ $y = $ _____	**c.** $5z = -35$ $z = $ _____

13. Write the equation and then solve it using "guess and check." Each root is between -20 and 20.

a. 2 plus 14 equals x minus 1
b. x cubed equals 27

14. Add or subtract.

| **a.** $(-9) + (-18) =$ _____ | **b.** $-21 - (-3) =$ _____ | **c.** $17 - 51 =$ _____ |

15. Give a real-life situation for the sum $3 + (-10)$.

16. Simplify.

    a. $|-2|$ **b.** $-(-2)$ **c.** $-|2|$ **d.** -0

17. Find the value of the expressions when $x = -2$ and $y = 8$.

| **a.** $5x^2$ | **b.** $-5y + 6$ | **c.** $-(y + x)$ |
| | | |

18. Jeremy is 2 years older than Larry. Write an expression for Larry's age, if Jeremy is y years old.

19. Here is a growing pattern. Draw the steps 4 and 5 and answer the questions.

Step   1   2   3

a. How do you see this pattern grow?

b. How many flowers will be in step 39?

c. In step n?

Mixed Review 4

1. Find the root(s) of the equation $x^2 - x - 20 = 0$ in the set $\{-5, -4, -3, 3, 4, 5\}$.

2. Is division commutative? Explain your reasoning.

3. Solve.

a. $\dfrac{x}{7} = -15$	**b.** $11 = \dfrac{x}{-12}$
c. $7 - x = -3$	**d.** $5 \cdot (-8) = -10x$

4. Interpret the absolute value in each situation.

   a. The temperature is $-4°C$. $|-4°C| =$ _____ $°C$

   Here, the absolute value shows _____

   b. The mountain is 2,500 ft tall. $|2,500 \text{ ft}| =$ _____ ft

   Here, the absolute value shows _____

5. Find the average speed in the given unit.

    a. Amanda swims 1 kilometer in 35 minutes.
      Give her average speed in kilometers per hour.

    b. You walk a distance of 1200 feet in 4 minutes.
      What is your average speed in miles per hour?

6. Add.

a. $(-3) + (-6) + 5 + 1 =$ _____	**b.** $14 + (-20) + (-31) + 11 =$ _____

7. Divide and simplify if possible.

a. $12 \div (-5)$	**b.** $-33 \div 15$	**c.** $-2 \div (-9)$

8. **a.** Jerry's yard is a rectangle with one 500-ft side and a total area of 150,000 square feet.
      How long is the other side? Write an equation with an unknown and solve it.

    b. Sketch a square that is one yard long and wide. Use your sketch to
      figure out how many square feet are in one square yard.

    c. Lastly, convert the area of Jerry's yard into square yards.

Rational Numbers Review

1. Write these numbers as a ratio (fraction) of two integers.

a. −3	b. 30	c. 0.21	d. −1.9

2. Mark the decimals on the number line: −0.21, −0.7, −0.03, −0.92

3. Write these decimals as fractions.

a. 0.0472	b. −1.02938442	c. 2.38166

4. Write the fractions as decimals.

a. $-\dfrac{24}{10,000}$	b. $\dfrac{9,872}{10}$	c. $\dfrac{4,593}{100,000}$

5. Write these repeating decimals using a horizontal line over the repeating part.

   a. 0.21212121…   b. 1.099555555…

6. Write these repeating decimals using an ellipsis (three periods).

   a. $2.06\overline{9}$ b. $0.006\overline{812}$

7. Which is more, 0.7, or $0.\overline{7}$?
   How much more?

8. Are all terminating decimals rational numbers?

   If not, give an example of a terminating decimal that is not a rational number.

9. Are all repeating decimals rational numbers?

   If not, give an example of a repeating decimal that is not a rational number.

10. Write as decimals. Calculate each answer to at least six decimal places. If you find a repeating pattern, then indicate the repeating part. If you don't, then round your answer to five decimals.

a. $\dfrac{3}{22}$

b. $1\dfrac{14}{23}$

11. The distance between two numbers a and b is given by the expression $|a - b|$. Show that this expression indeed gives the distance between the two given values of a and b by evaluating the expression and by also calculating the distance using logical reasoning.

a. $a = 6$ and $b = -7$

Distance:

Absolute value of the difference:

$|\ \boxed{}\ -\ \boxed{}\ | =$

b. $a = -1.3$ and $b = -7.6$

Distance:

Absolute value of the difference:

$|\ \boxed{}\ -\ \boxed{}\ | =$

12. Multiply mentally.

a. $0.2 \cdot 0.07$	**b.** $-0.8 \cdot 0.005$	**c.** $(-0.2)^3$
d. $-5 \cdot (-2.2)$	**e.** $-0.2 \cdot 0.1 \cdot (-0.3)$	

13. Multiply

a. $-\dfrac{4}{11} \cdot \left(-\dfrac{7}{12}\right)$	b. $\dfrac{5}{6} \cdot \left(-3\dfrac{1}{2}\right)$	c. $-\dfrac{9}{20} \cdot \dfrac{2}{3} \cdot \left(-\dfrac{1}{5}\right)$

14. Divide.

a. $1\dfrac{1}{5} \div \left(-\dfrac{1}{4}\right)$	b. $21 \div 0.06$

15. Solve *without* a calculator.

a. 60% of $18	b. $\dfrac{1}{4} \cdot 9.6$	c. $-0.3 \cdot \dfrac{8}{11}$

16. Simplify these complex fractions.

a. $\dfrac{3}{\frac{2}{5}}$	b. $\dfrac{\frac{6}{7}}{\frac{5}{12}}$	c. $\dfrac{\frac{8}{3}}{2}$

17. Give a real-life context for each calculation. Then solve.

a. $1.56 \cdot 0.8$

b. $6 \div (1/2)$

18. A **kilowatt-hour** (kWh) is a unit of energy. It describes the energy consumed (by some device) that uses one **kilowatt** (kW) of power for one hour. The formula is: **power × time = energy**. So you multiply the kilowatts times the hours to get kilowatt-hours. For example, let's say a 2kW air conditioner runs for one hour. Then it uses 2 kWh (two kilowatt-hours) of energy in that time. Your electric company charges you for the amount of *energy* that you consume.

If electricity costs 16.86 cents per kWh, then how much would it cost to run a 2kW air conditioner for 16 hours each day during June, July, and August?
Hint: First calculate how much energy (in kilowatt-hours) the AC unit uses in that time period.

19. Write the numbers in scientific notation.

a. 6,798,000

b. 56,000,000,000

20. Write the numbers in numerical form.

a. $7.8 \cdot 10^5$

b. $3.4958 \cdot 10^9$

21. Solve the problem using an equation and also using some other strategy.

| A forest fire was 7/10 contained. The contained area was 4,200 acres. What was the total area of the fire? |
| Equation: |

Another way:

22. Solve.

a. $x - \dfrac{2}{9} = 5\dfrac{1}{20}$

b. $5y = -\dfrac{4}{12}$

c. $0.94 = 1.1 - x$

d. $-0.3x = 10$

Rational Numbers Test

1. Mark these numbers on the number line: $-\dfrac{2}{5}$, $-1\dfrac{4}{5}$, $-2\dfrac{1}{5}$, $-\dfrac{1}{10}$, $-1\dfrac{9}{10}$

```
     |----+----+----+----+----+----+----+----+----+----+----|
        -2            -1             0             1
```

2. Form a fraction (numerator and denominator) from the two given integers. Then give it as a decimal.

a. 5 and 4	**b.** −7 and 10	**c.** 9 and −100

3. Write the decimals as mixed numbers.

a. 5.001	**b.** −2.0482

4. Write the fractions as decimals.

a. $-\dfrac{47}{10,000}$	**b.** $\dfrac{787}{10}$	**c.** $-\dfrac{5,624}{100}$

5. Which is more, $0.\overline{6}$ or 0.6?
   How much more?

6. Write as decimals, using a line over the repeating part (if any). Use long division.

a. $\dfrac{7}{6}$	**b.** $\dfrac{5}{36}$

7. Add or subtract.

a. $-1.26 - (-3.45)$	b. $1.8 - 3.25$	c. $-0.42 + 10.7 + (-9.8)$

8. Add or subtract.

a. $\dfrac{5}{9} + \left(-\dfrac{2}{3}\right)$	b. $-\dfrac{1}{10} - \dfrac{6}{9}$

9. Write the numbers in scientific notation.

a. 25,600,000	b. 7,810,000,000

10. Multiply or divide.

a. $-0.06 \cdot 0.05$	b. $(-0.5)^3$
c. $\dfrac{1}{3} \cdot \left(-5\dfrac{6}{11}\right)$	d. $-6\dfrac{1}{9} \div \left(-\dfrac{3}{4}\right)$

11. Divide *without* a calculator.

a. $1.5 \div 0.006$	b. $0.9 \div 0.011$

12. Give a real-life context for the calculation $\dfrac{1}{3} \cdot 12.75$. Then solve.

13. Find 15% of 3/4.

14. Two-thirds of a number is −5.66.
    What is the number?

15. Simplify these complex fractions.

a. $\dfrac{\frac{2}{5}}{4}$	b. $\dfrac{\frac{9}{10}}{\frac{1}{6}}$

Mixed Review 5

1. Jeremy purchased x pairs of gloves for $3 each and one pair of rubber boots for $9.

   a. Write an expression for the total cost of his purchases.

   b. The total cost of Jeremy's purchases was $57.
   Write an equation for this situation.

   c. How many pairs of gloves did he buy?
   You can solve the equation or figure it out using mental math.

2. Solve. Simplify one side first.

a. $\quad 2r - 5 \;\; = \;\; 10 - (-2)$	**b.** $\quad 2 \cdot 3 \;\; = \;\; 9 - 6y$

3. Light travels at a speed of 299,792.458 kilometers per second. That is quite fast!

   a. A beam of light is sent from the earth to the moon. How long will it take to reach the moon?
   The average distance between the earth and the moon is 384,403 km.
   Give your answer to five decimal digits.

   b. (Optional.) Choose some other object in our solar system, and calculate how long it will take a radio
   message sent from the earth to reach that object. Radio waves travel at the speed of light.
   Use an encyclopedia to find the distances you need to solve the problem.

4. Solve. Check your solutions.

a. $\quad 2 - x \;=\; -6$	**b.** $\quad -10 - x \;=\; 7$
c. $\quad 2x \;=\; -5$	**d.** $\quad 2 + (-11) \;=\; 8 + z$

5. Divide and simplify if possible.

a. $1 \div (-3)$	**b.** $-16 \div 20$	**c.** $-45 \div (-36)$

6. Write an addition or subtraction using integers to match the situation.

   a. A shark was swimming at the depth of 4 m. Then it rose 2 m.

   b. Michael owed $250. He then purchased a computer on his credit
   card for $500 (adding to his debt).

7. Solve. Check your solutions.

a. $\quad \dfrac{x}{-13} \;=\; 4$	**b.** $\quad \dfrac{w}{-3} \;=\; -11 + 5$
c. $\quad -31 \;=\; \dfrac{1}{6}x$	**d.** $\quad 1 \;=\; -5x$

Mixed Review 6

1. Are the expressions equal, no matter what values n and m have? If so, you don't need to do anything else. If not, provide a counterexample: specific values of n and m that show the expressions do NOT have the same value.

a. $(-n-1)-m$ $-n-(1-m)$	**b.** $\dfrac{x-y}{2}$ $\dfrac{x}{2} - \dfrac{y}{2}$

2. **a.** Sketch a rectangle with sides $3s$ and $8s$ long.

   b. What is its area?

   c. What is its perimeter?

3. Simplify the expressions.

a. $23r - 8r + 7r + 5$	**b.** $9p^2 + 8 - 3p^2$	**c.** $6y \cdot y \cdot 7y$

4. **a.** What is the total value, in cents, if Ashley has n quarters?
   Write an expression.

   b. Let's say the total value of her quarters is 875 cents.
   How many quarters does Ashley have?
   Write an equation and solve it.

5. Factor these sums (write them as products).

a. $100x + 60 =$	**b.** $24s - 4t - 8 =$

6. Consider the four expressions $67 + 28$, $(-67) + (-28)$, $(-67) + 28$, and $67 + (-28)$. Write these expressions in order from the one with **least** value to the one with **greatest** value.

7. **a.** Find the value of the expression $3 - x$ for at least six different values of x. Make a table to organize your work. Choose values of x in a pattern and notice the pattern in the values of $3 - x$.

   b. For which value of x will the expression $3 - x$ have a value of -2?

8. Solve the equation using the balance model. Write in the margin what operation you perform on both sides.

Balance	Equation	Operation to perform on both sides
	$4x - 1 = 7$	

Equations and Inequalities Review

1. Solve. Check your solutions (as always!).

a. $\quad 1 - 3x \;=\; 17$	**b.** $\quad 29 \;=\; -6 - 2y$
c. $\quad \dfrac{3x}{8} \;=\; 42$	**d.** $\quad \dfrac{v-2}{7} \;=\; -13$
e. $\quad \dfrac{w}{40} - 7 \;=\; 19$	**f.** $\quad \dfrac{s+8}{-3} \;=\; -1$

2. Solve each problem in two ways: (1) by writing an equation and (2) by using logical reasoning or a bar model.

a. You bought 15 bottles of oil for the equipment in your lawn-care business. You got a $14 discount on your entire purchase. The total cost was $130 after the discount. What is the normal price of one bottle of oil?

Equation:

Logical thinking:

b. Three-sevenths of a number is 153. What is the number?

Equation:

Logical thinking:

3. A carpet salesman earns a base salary of $300 a week. He also earns an additional $18 for every carpet he sells.

    a. Write an expression for the salesman's total weekly earnings if he sells n carpets.

    b. How many carpets does the salesman need to sell in order to earn $750 in a week? Write an equation and solve it.

4. Solve. Check your solutions.

a. $\quad 2x + 6 + 3x \;=\; 9x - 11$	**b.** $\quad 2(x + 6) \;=\; 9x - 11$
c. $\quad 6(5 - w) \;=\; 2(9 - w)$	**d.** $\quad -10(4y + 7) \;=\; -9y$

5. Four adjacent (side-by-side) angles form the line *l*.

    a. Write an equation to solve for the unknown *x*.

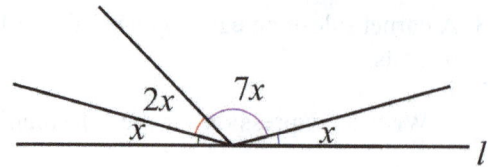

    b. Solve your equation and find the measure of each of the four angles.

6. The total area of a divided room is 200 square feet. Find the unknown dimension.

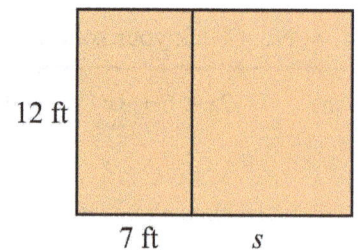

12 ft

7 ft   *s*

7. Solve the inequalities and plot their solution sets on the number line. You need to write appropriate numbers for the tick marks yourself.

a. $5x - 8 \; < \; 22$	**b.** $x + 5 \; \geq \; -2$

8. Write an equation for this number line diagram and solve it to find the value of the unknown y.

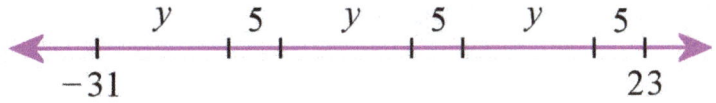

9. An airline has a weight limit of 20 kg for carry-on bags (the luggage passengers carry onto the airplane). Sharon's clothes, personal items, and the carry-on bag itself weigh 9 kg. Besides those, she wants to take a camera that weighs 2.6 kg and as many 0.8-kg bags of nuts as she can. How many bags of nuts can she take?

a. Solve the problem without an equation or inequality.

b. Write an inequality for the problem and solve it.

10. Find the slope of each line. Also, graph the lines.

a. $y = -2x - 1$

x							
y							

Slope: _____

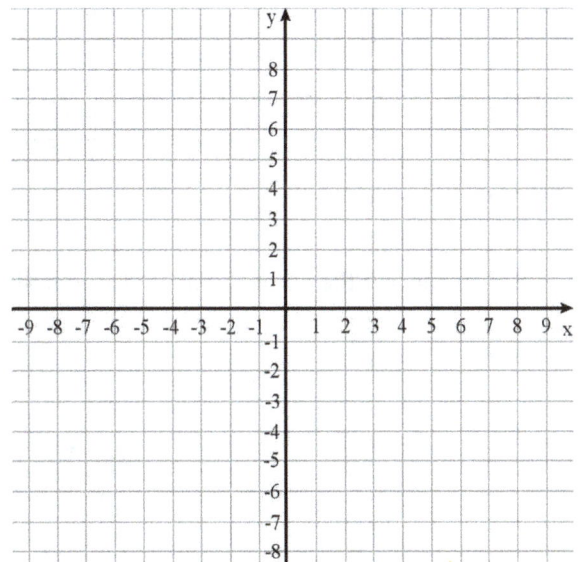

b.

x	−3	−2	−1	0	1	2	3
y	−6 ½	−4	−1 ½	1	3 ½	6	8 ½

Slope: _____

45

11. Draw a line with a slope of 5/6.

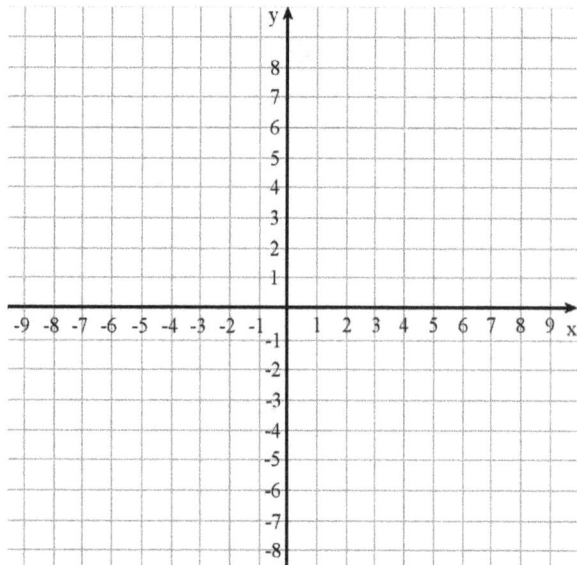

12. Draw a line that has a slope of 3/2 and that goes through the point (0, 2).

13. An airplane travels at a constant speed of 600 mi/h from New York to Los Angeles, a distance of 2,450 miles.

a. Write an equation relating the distance (d) it has traveled and the time (t) that has passed.

b. Plot your equation. Notice that you need to scale the d-axis.

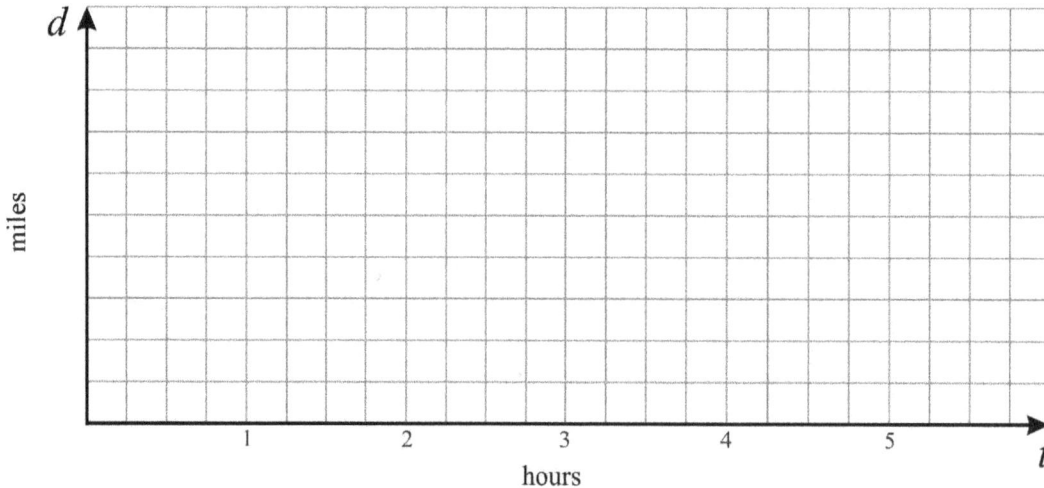

c. How far will the airplane travel in 1 hour 40 minutes?

Equations and Inequalities Test

1. Solve. Give your answers as fractions, mixed numbers, or whole numbers (not decimals).

a. $-2 = 6x + 5$

b. $6x + 2x - 1 = -9x + 1$

c. $\dfrac{3x}{5} = 24$

d. $\dfrac{y}{3} - 21 = -5$

2. Ethan purchased 24 cookies and a loaf of bread for a total of $6.85. He didn't pay attention to the cost of the cookies but he remembered that the bread cost $3.25. Find the cost of one cookie by writing an equation and solving it.

3. Solve the inequalities and plot their solution sets on a number line. Write appropriate numbers for the tick marks yourself.

a. $3x + 5 < 68$	b. $10x - 17 \geq 103$

4. Abner got a building permit for a shed that limits its height to a maximum of 9.5 ft above the surface of the ground. He poured an above-ground foundation that was 4 inches thick, and the flat roof will add 6 inches to the height of the wall. He is going to build with concrete blocks that are 8 inches high (including the mortar).

   a. Write an inequality to calculate how many rows of block he can lay without the shed exceeding the maximum height permitted.

   b. Solve the inequality.

   c. Draw a number line and plot the solution set.

5. Plot the equations.

    a. $y = x + 4$

    b. $y = -3x + 5$

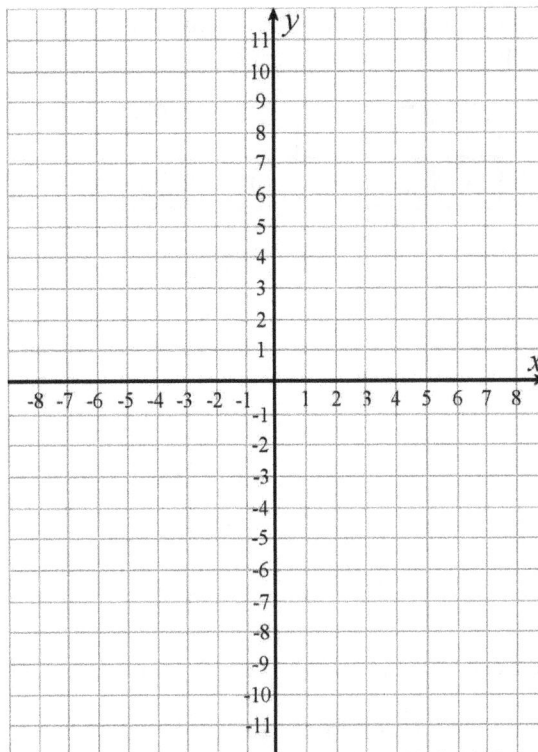

6. **a.** Draw any line that has a slope of −2.

    b. Draw a line that has a slope of 1/2 and
        that goes through the point (1, 3).

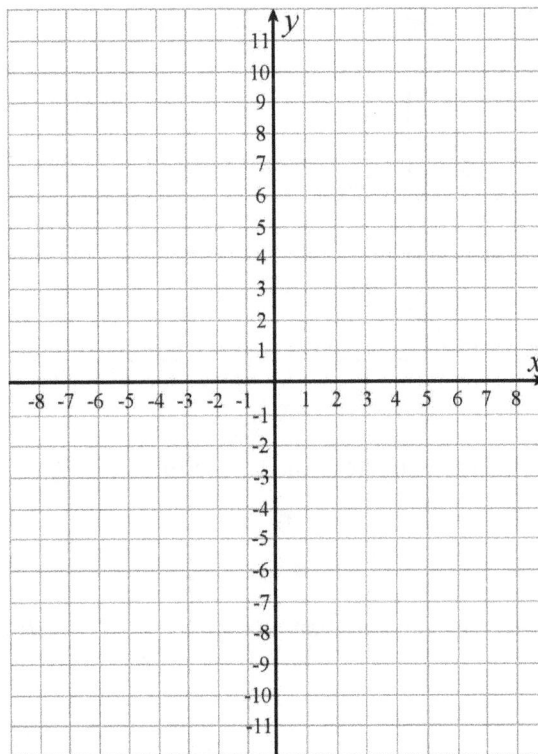

7. Determine the slope of this line from
   the table or from its graph.

x	0	1	2	3	4	5	6	7
y	20	30	40	50	60	70	80	90

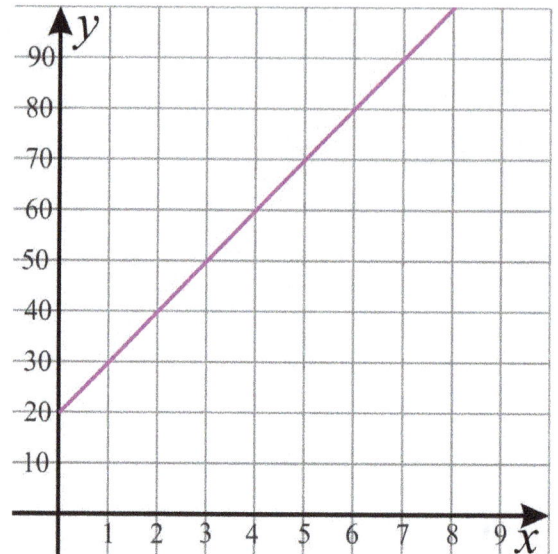

8. Leah runs at a constant speed of 4 m/s for 100 meters.

   a. Plot a graph for the distance Leah runs.

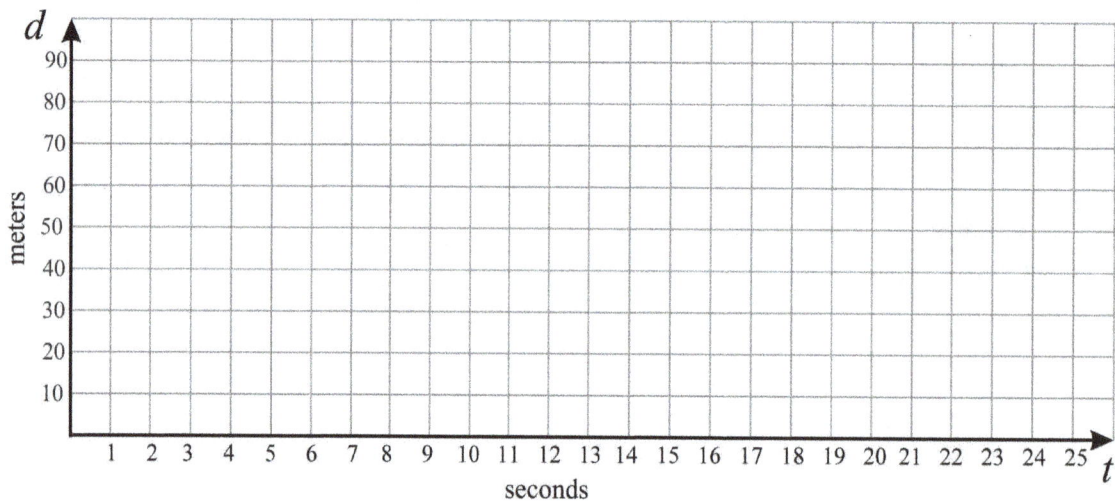

   b. How long does she take to run the 100 meters?

   c. Write an equation relating the distance (*d*) Leah has run and the time (*t*) that has passed.

50

Mixed Review 7

1. Find the missing numbers and terms.

a. _____ $(6x - 5) = 72x - 60$	**b.** $12(\_\_\_\_ - \_\_\_\_ + \_\_\_\_) = 108y - 36x + 4.8$

2. Write an expression with two terms: the coefficient of the first term is 5,
   its variable part is x cubed, and the second term is the constant $-1/2$.

3. **a.** Which equation matches the situation?

   A town of population p lost 2/3 of its population, and now it has 2,600 residents.

$p - 2/3 = 2600$	$\dfrac{2p}{3} = 2600$	$p - 1/3 = 2600$	$p - 2/3p = 2600$

   b. How many people lived in the town originally?

4. Add.

a. $(-14) + 7 + (-8) + 2 =$	**b.** $-3 + (-12) + 21 + (-19) + (-5) =$

5. Give a real-life context for each multiplication. Then solve.

a. $1.4 \cdot 119$
b. $(9/10) \cdot 14.30$

6. Change each subtraction into an addition, then add.

a. $-8 - (-7) - (-12) =$	b. $63 - (-11) + (-5) =$

7. **a.** Write an expression for the distance between x and 8.

   b. Evaluate your expression if $x = -52$.

8. Solve using both decimal and fraction arithmetic.

a. $0.24 \div 0.03$ **Decimal division:**   **Fraction division:**
b. $7.1 \cdot 0.5$ **Decimal multiplication:**   **Fraction multiplication:**

9. Solve.

$$\frac{5}{6} \cdot \frac{2}{3} \div \frac{4}{3}$$

10. Solve *without* a calculator.

a. 11% of $15	b. 90% of −12	c. 75% of −200 m

11. Solve *without* a calculator. Change the decimals into fractions or treat fractions as divisions.

a. $0.5 \cdot \frac{11}{12}$	b. $\frac{2}{5} \cdot (-0.8)$	c. $-\frac{5}{6} \cdot 0.2$

12. Rewrite each expression without parentheses.

a. $2 + (-g) =$	b. $15 - (-r) =$	c. $7x + (-2y) =$

13. Write the numbers in scientific notation.

    a. 113,000   b. 45,980,000

14. Simplify the complex fractions.

a. $\dfrac{\frac{7}{8}}{\frac{8}{9}}$	b. $\dfrac{\frac{1}{2}}{\frac{1}{5}}$	c. $\dfrac{\frac{15}{21}}{\frac{2}{3}}$

Mixed Review 8

1. **a.** Which expression can be used to find the distance between x and 6?

(i) $|x - (-6)|$	**(ii)** $x - (-6)$	**(iii)** $x - 6$	**(iv)** $|x - 6|$	**(v)** $|x + 6|$

   b. Evaluate the expression when x is -23.

2. Justify the rule "A negative times a negative makes a positive" by filling in the missing parts of this proof based on the distributive property.

   (1) Substitute $a = -1$, $b = 1$, and $c = -1$ into the formula for the distributive property $a(b + c) = ab + ac$.

   _____ (_____ + _____) = _____ · _____ + _____ · _____

   (2) The whole left side is zero because _____ + _____ = 0.

   (3) So the right side must equal zero as well.

   (4) On the right side, $-1 \cdot 1$ equals _____ . So, $-1 \cdot (-1)$ must equal _____ so that the sum on the right side will equal zero.

   (5) Therefore, $-1 \cdot (-1)$ must equal _____ .

3. Multiply.

a. $(-7) \cdot 2 \cdot (-2)$	**b.** $10 \cdot (-4) \cdot 7$	**c.** $2 \cdot (-5) \cdot (-2) \cdot (-5)$

4. Below you see listed the minimum daily temperatures for one week. Calculate their average.

   $-8°C, -11°C, 2°C, 0°C, -3°C, -5°C, -1°C$

5. Multiply mentally.

a. $0.3 \cdot 2.5$	**b.** $-0.002 \cdot 0.008$	**c.** $-0.9 \cdot 50$
d. 0.8^2	**e.** $-4 \cdot 0.05 \cdot (-20)$	**f.** $(-0.3)^2$

6. Write the fractions as decimals.

a. $-\dfrac{61}{100,000}$	b. $\dfrac{9,807,200}{1000}$	c. $\dfrac{55,191}{1,000,000}$

7. Alex commutes 12 km to work every day. One day, his average speed going to work was 60 km/h and coming back 50 km/h. How long did it take Alex to commute that day?

8. Write in decimal form. Use long division, and calculate each answer to at least six decimal places. If you find a repeating pattern, give the repeating part. If you don't, round your answer to five decimals.

a. $2\dfrac{5}{24}$	b. $2.05 \div 7$	c. $5.6 \div 0.02$

9. Add the fractions.

a. $\dfrac{3}{5} + \left(-\dfrac{2}{3}\right)$	b. $-\dfrac{1}{2} + \left(-\dfrac{6}{9}\right)$

Ratios and Proportions Review

1. Simplify the ratios and rates.

a. 164 km per 4 hours	**b.** $\dfrac{6 \text{ g}}{1600 \text{ ml}} =$	**c.** 52 : 156 = _____ : _____

2. A car traveled 348 miles in 6 hours. Fill in the table of equivalent rates.

Miles						348		
Hours	1	2	3	4	5	6	10	20

3. A mixture of salt and water contains 20 grams of salt and 1,200 grams of water.
   Write the ratio by weight of salt to water and simplify it.

4. Susan can jog 1 1/2 miles in 1/3 hour.
   Write a rate for her jogging speed and simplify it.

5. Solve the proportions. Round your answers to the nearest hundredth.

a. $\dfrac{16}{17} = \dfrac{109}{T}$	**b.** $\dfrac{1.5}{2.8} = \dfrac{M}{5}$

6. Write a proportion for the following problem and solve it.

12 kg of chicken feed costs $19.
How much would 5 kg cost? $\underline{\hspace{3cm}} = \underline{\hspace{3cm}}$

7. On average, Gary makes a basket eight times out of every ten shots.
How many baskets can he expect to make when he practices 25 shots?

8. Write the unit rate as a complex fraction, and then simplify it.

a. Alex solved 2 1/2 pages of math problems in 1 1/4 hour.

b. Noah painted 2/3 of a room in 3/4 of an hour.

9. A car is traveling at a constant speed of 75 km per hour.

   a. Write an equation relating the distance (d) and time (t) and plot it in the grid below.

   b. What is the unit rate?

   c. Plot the point that matches the unit rate in this situation.

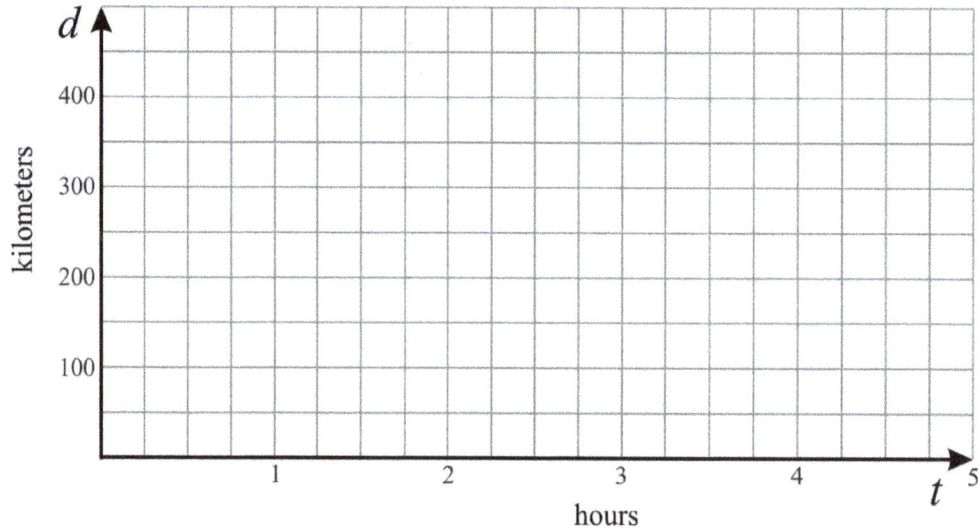

   d. What does the point (0, 0) mean in terms of this situation?

   e. How far can the car travel in 55 minutes, driving at the same speed?
   Also, plot the point for the time $t = 55$ min.

   f. How long will the car take to travel 160 km? Give your answer in hours and minutes.
   Also, plot the point that matches the distance $d = 160$ km.

10. Using a pre-paid internet service you get a certain amount of bandwidth to use for the amount you pay. The table shows the prices for certain amounts of bandwidth.

Bandwidth	1G	2G	5G	10G	15G	20G	25G
Price	$10	$16	$23	$30	$37	$43	$50

a. Are these two quantities in proportion?

Explain how you can tell that.

b. If so, write an equation relating the two and state the constant of proportionality.

11. In the year 2008 it was estimated that it cost $9,369 a year to drive a medium-sized car (a sedan) for 15,000 miles (a typical amount of use). Based on those same assumptions, how much would it cost, to the nearest dollar, to drive that car for 5 months?

12. The figures are similar. Find the length of the side labeled with x.

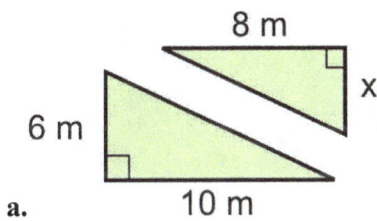

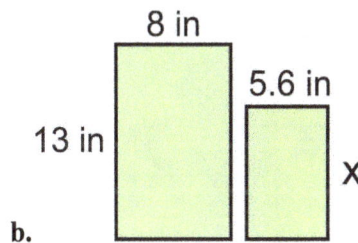

a.

8 m

6 m

x

10 m

b.

8 in

5.6 in

13 in

x

13. A house plan has a scale of 1 in : 6 ft. In the plan, one room measures 2 in × 2 ¾ in. What are the true dimensions of the room?

14. A freight truck fully loaded with cargo gets six miles to a gallon of diesel.

   a. What is the unit rate in this situation?

   b. Write an equation relating the mileage (M) to the amount of diesel fuel (f) in gallons.

   c. Plot your equation. Choose an appropriate scaling for the two axes.

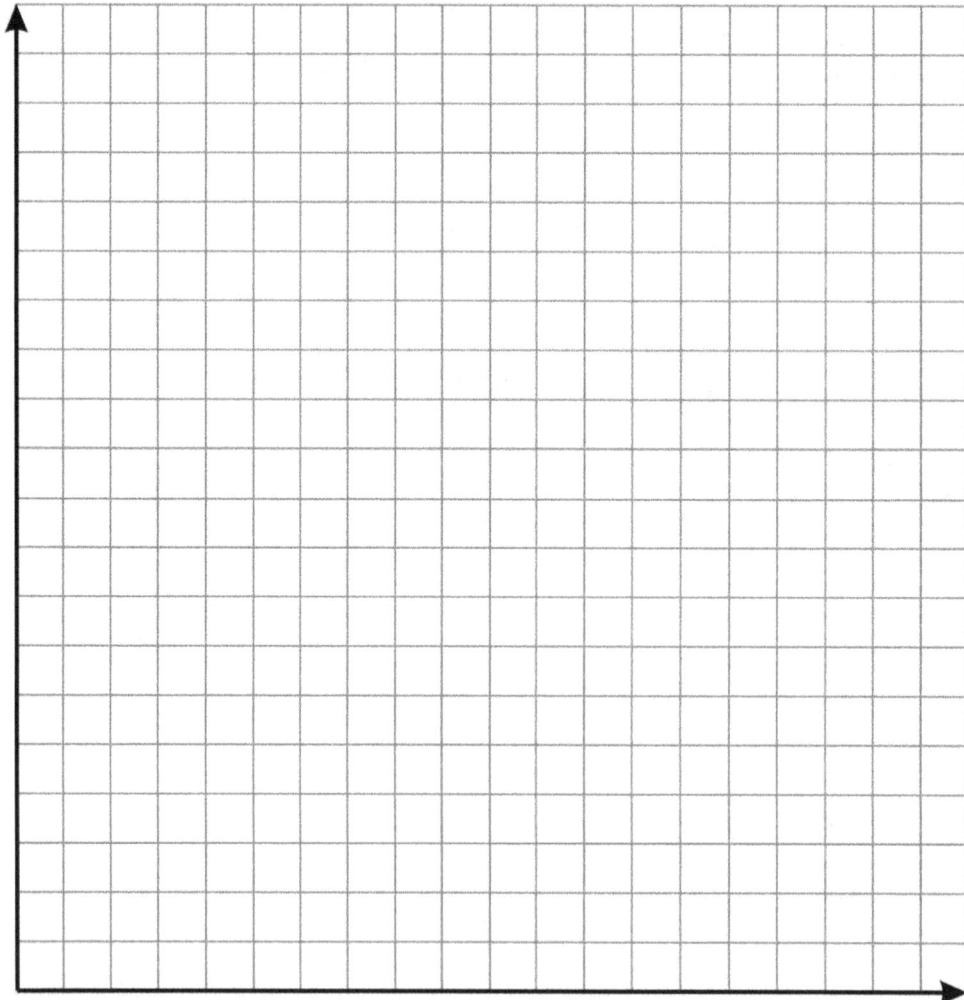

Ratios and Proportions Test

You may use a calculator for all the problems in this test.

1. Chloe bicycled 20 kilometers in 1 1/2 hours.
   Write a rate for her speed and simplify it to find the unit rate.

2. Mason poured 1/3 of an envelope of chocolate drink powder into 2/3 cups of water.

   a. Write the unit rate as a complex fraction and simplify it.

   b. What does the unit rate signify?

3. Write a proportion for the following problem and solve it.

   A bag of 52 kg of wheat costs $169.
   What would 21 kg of wheat cost?   ——————— = ———————

4. Solve the proportion by using cross-multiplication.

$$\frac{4.3}{S} = \frac{7.9}{12}$$

5. The figures are similar. Find the length of the side labeled with x.

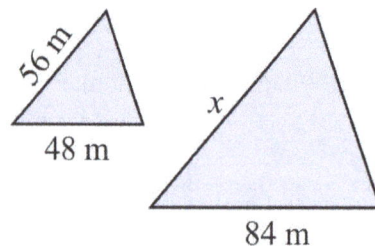

56 m

48 m

x

84 m

6. The aspect ratio of a television screen is 16:9 (width to height), and it is 63 cm high. What is its width?

7. A town map has a scale of 1:45,000.

a. A street in this town is 850 m long. How long is that street on this map?

b. How long in reality is a road that measures 5.4 cm on the map?

8. The graph on the right depicts the distance that a running fox covers as time passes.

   a. State the unit rate (including the units of measurement) for this situation.

   b. Plot the point that corresponds to the unit rate.

   c. Write an equation for the line.

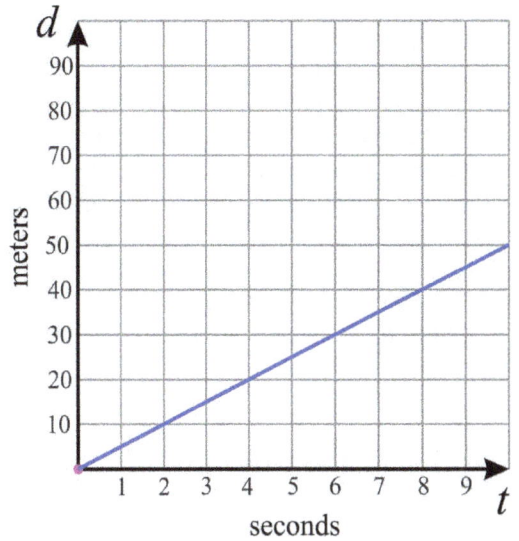

9. The equation S = 15t tells us the salary (S, in dollars) of a worker who works for t hours.

   a. What is the unit rate in this situation?

   b. Graph the equation S = 15t in the grid. Label the axes. Choose the scaling for the two axes so that the point corresponding for working for 10 hours will fit on the grid.

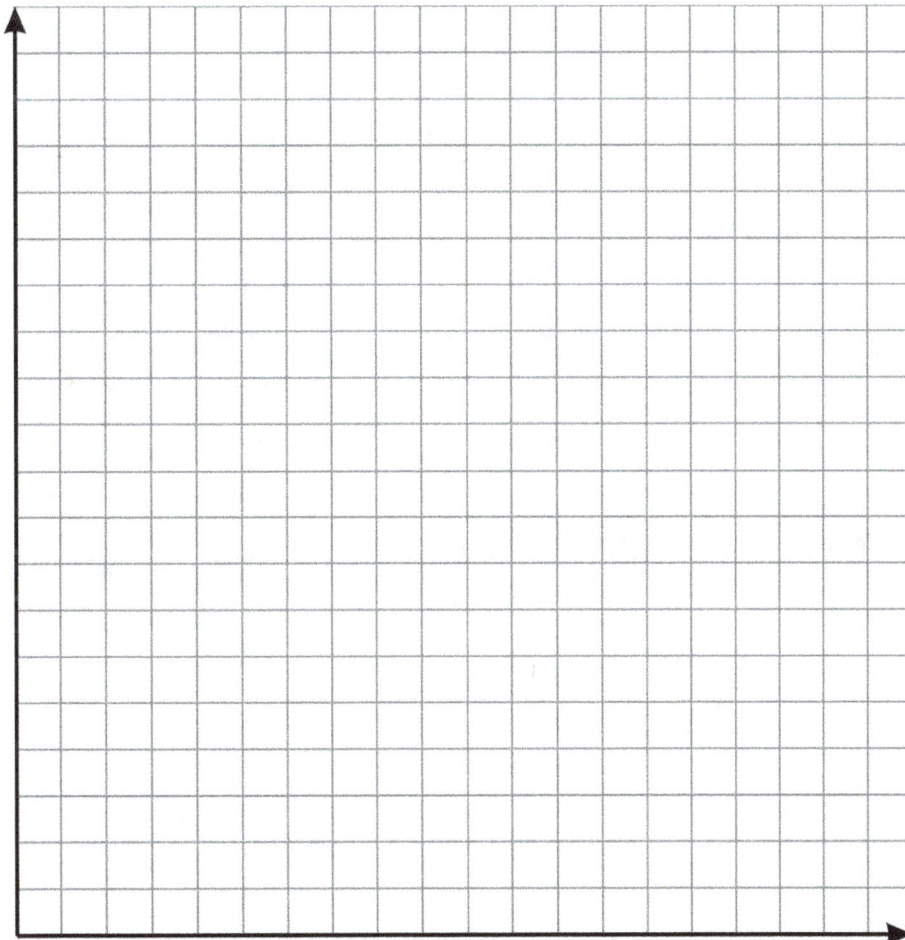

Mixed Review 9

1. The table lists the average high temperatures for January and July for several Canadian cities. It gives you an idea of how warm it gets during the winter and during the summer, on average.

City	January (Avg. High °C)	July (Avg. High °C)
Winnipeg, MB	−11.9	25.9
Saskatoon, SK	−10.1	25.3
Quebec City, QC	−7.0	24.7
Edmonton, AB	−6.3	22.8
Ottawa, ON	−5.8	26.6
Calgary, AB	−0.9	23.2
Montreal, QC	−5.3	26.3
Halifax, NS	−0.1	23.1
Toronto, ON	−0.7	26.6
Vancouver, BC	6.8	22.1
Yellowknife, NT	−21.6	21.3
Iqaluit, NU	−22.8	12.3

   a. In which city is the difference between the average high temperatures in these two months the greatest?

   b. How much is that difference?

   c. In which city is the difference the smallest?

   d. How much is that difference?

2. Write the numbers in scientific notation.

   a. 2,089,000

   b. 394,410,000

3. Divide and simplify.

a. $2 \div (-8) = -\dfrac{2}{8} = -\dfrac{1}{4}$	**b.** $4 \div (-24)$	**c.** $-18 \div 5$
d. $-2 \div (-9)$	**e.** $42 \div (-49)$	**f.** $-32 \div (-28)$

4. Find the value of the expressions using the correct order of operations.

a. $5 \cdot \dfrac{2}{-10}$	**b.** $-\dfrac{12}{-4} + 7$	**c.** $-1 + \dfrac{24}{12 + (-6)}$
d. $-2 + 7 \cdot 2 - 6$	**e.** $-8 \cdot (-7) - 11$	**f.** $(-3 + 9) \cdot 8$

5. Solve. Check your solutions (as always!).

a. $11 - 5x = -6$

Check:

b. $6(y + 2) = -16$

Check:

c. $\dfrac{2x}{5} = 30$

Check:

d. $\dfrac{s - 12}{5} = -1$

Check:

6. Find the slope of the line.

7. Write an inequality for each situation. Use a variable for the quantity in question (the temperature, the amount of flour, and the cost).

   a. The temperature shouldn't exceed 42°C.

   b. I need at least 3 cups of flour.

   c. The cost has to be kept strictly under $3,000.

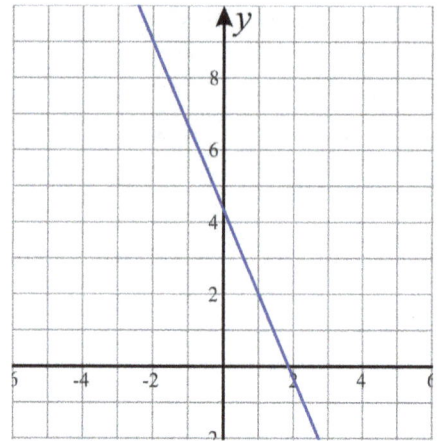

8. A farmer is hiring workers to help him on his farm. He will pay each worker a monthly salary of $2,050. The farmer has a total budget of $40,000 for three months. How many workers can he hire?

   a. Solve the problem without using an equation or inequality.

   b. Write an inequality for the problem, and solve it.

9. **a.** Plot the equation $y = (3/4)x$.

   b. Plot the equation $y = -2x + 1$.

   c. Plot the equation $y = 4 - x$.

10. Determine whether the point $(2, -3)$ is on the line $y = -2x - 2$. Justify your answer.

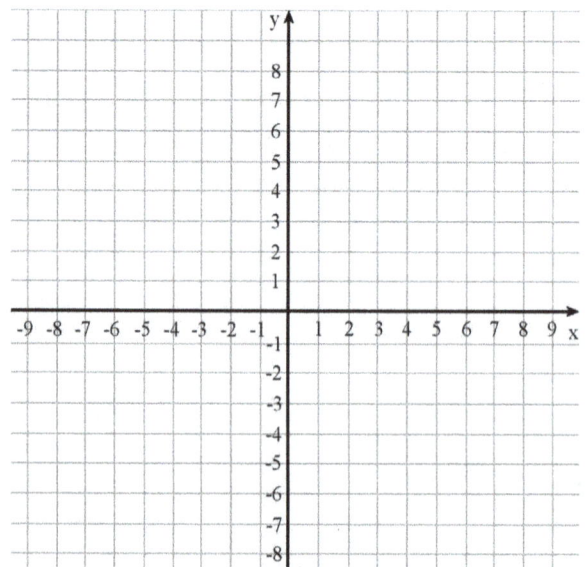

Mixed Review 10

1. Simplify the expressions.

a. $-3z - 9 + 7z + 2t$	**b.** $6x \cdot x \cdot (-7x)$	**c.** $6s \cdot s \cdot 4t$

2. Solve *without* a calculator.

a. 30% of $400	**b.** $\dfrac{2}{3} \cdot 6.9$ km	**c.** $0.08 \cdot \dfrac{1}{10}$

3. Solve. Check your solutions.

a. $2x - 6 = 9x - 8$	**b.** $3(x - 6) = -9x$
c. $8x = -\dfrac{3}{4}$	**d.** $1\dfrac{1}{6} + v = \dfrac{2}{9}$

4. Find the slope of each line.

   a.

   b.

   c.

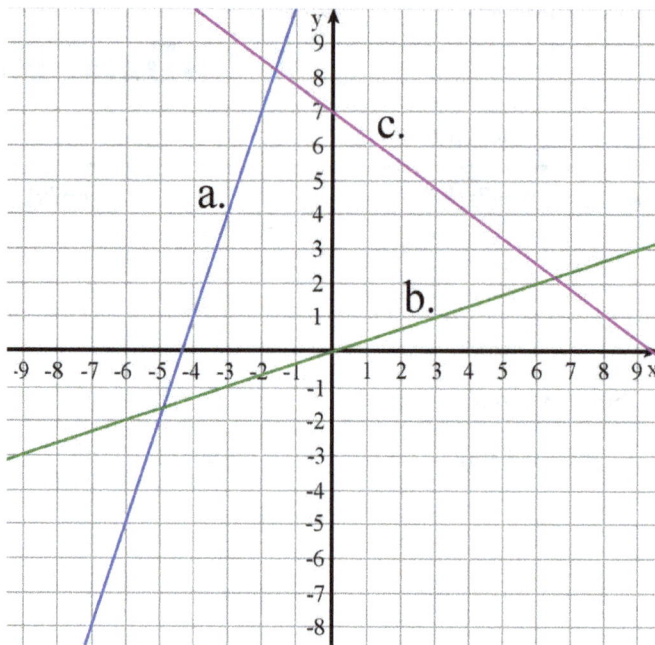

5. Write the numbers in scientific notation.

   a. 490,000,000

   b. 6,238,000,000

6. Write the numbers in numerical form.

   a. $2.08 \cdot 10^8$

   b. $1.293 \cdot 10^6$

7.**a.** Draw a line that has a slope of 2 and
      that goes through the point (0, 6).

   b. Draw a line that has a slope of −1/2 and
      that goes through the point (−4, 7).

   c. Draw a line that has a slope of 4/3 and
      that goes through the point (0, 1).

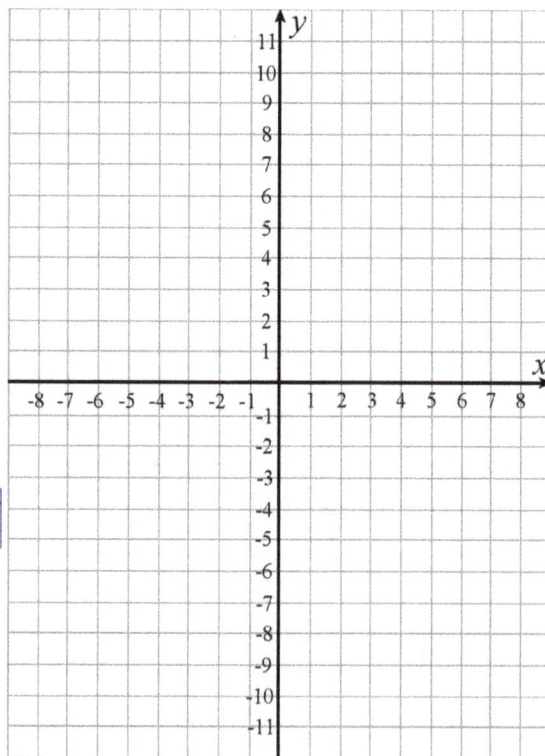

8. Cynthia took 14 minutes to bicycle from her home to
   a dentist appointment (a distance of 2.8 km) and
   10 minutes to bicycle from there back home.
   Calculate her average speed for the entire trip.

9. A contractor has quoted you a price of $4.50 per square foot for a driveway. The driveway will be 40 ft long but you need to decide the exact width. You can afford to spend at most $1,600.

    a. Find the maximum width for the driveway.
      Hint: first find the maximum area for the driveway based on what you can afford.

    b. Write an equation for finding the maximum width of the driveway, and solve it. I realize you already know the answer from solving it in (a), but the purpose of this exercise is to let you practice how to write equations for real-life situations.

10. An airplane travels at a constant speed of 500 mi/h.

    a. Write an equation relating the distance (d) it has traveled and the time (t) that has passed.

    b. Plot your equation. Notice that you need to scale the d-axis.

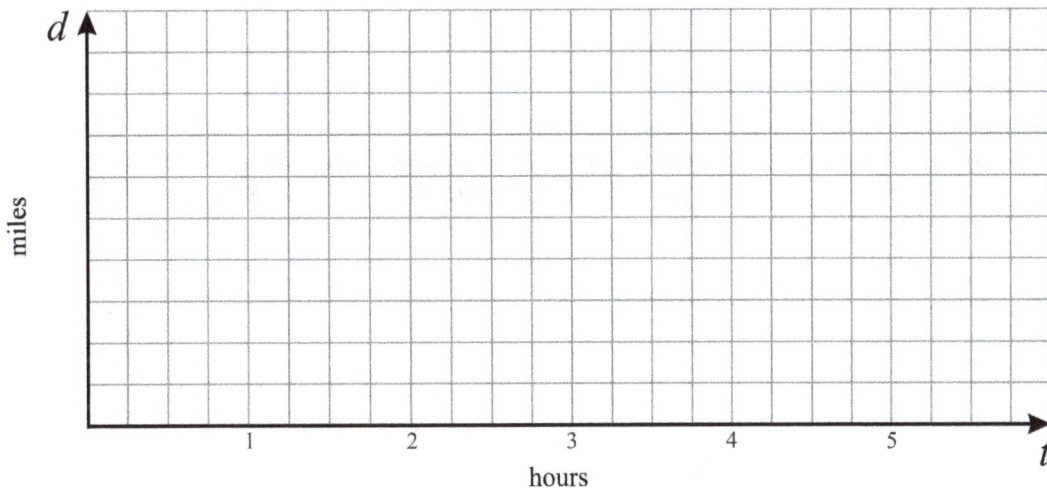

    c. How long will it take the airplane to travel 3,600 miles?

Percent Review

1. Find the percentage of the given quantity.

   a. 9.2% of $150   **b.** 45.8% of 16 m   **c.** 0.6% of 700 mi

2. All these items are on sale. Calculate the new, discounted prices.

   a.

   Price: $9
   20% off

   New price: $_____

   b.

   Price: $6
   25% off

   New price: $_____

   c.

   Price: $90
   30% off

   New price: $_____

3. A flashlight is discounted by 18%, and now it costs $23.37. Let p be its price
   before the discount. Find the equation that matches the statement above
   and solve it.

   $p - 0.18 = \$23.37$

   $p - 18 = \$23.37$

   $0.82p = \$23.37$

   $0.18p = \$23.37$

4. Two brothers, Andy and Jack, shared the price of a new computer so that Andy paid
   2/5 (or 40%) and Jack paid 3/5 (or 60%) of the price. The computer cost $459, and there
   was a sales tax of 7% that was added onto the price. Calculate Andy's and Jack's shares.

5. A portable reading device costs $180. Now it is discounted and costs $153.
   What was the percentage of discount?

6. The two right triangles on the right are similar.

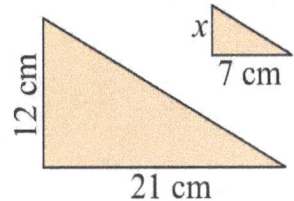

   a. Calculate what percentage the area of the smaller triangle
      is of the area of the larger triangle.

   b. In what ratio are the corresponding sides of the triangles?

   c. In what ratio are their areas?

7. A wall painting was planned to be 5 m by 3 m in size. If both of its sides are increased by 20%,
   by what percentage will the area of the painting increase?

8. In a race, Old Paint finished in 120 seconds, and The Old Gray Mare finished in 200 seconds.

   a. How many percent quicker was Old Paint than The Old Gray Mare?

   b. How many percent slower was The Old Gray Mare than Old Paint?

   c. What is the relative difference in their times?

9. Noah takes a $4,000 loan at a 7.8% annual interest rate to purchase a car. At the end of the first year, he
   pays back $2,000 of the principal of the loan. At the end of the second year, he pays back the rest of the
   principal and all of the interest. How much total interest does he have to pay? Assume simple interest,
   which is calculated and paid only at the end of the period of the loan.

Percent Test

You may use a basic calculator for all the problems in this test.
If not otherwise specified, give your answers that are percentages to the tenth of a percent.

1. The price of these items is changing. Find the new price or the discount percentage.

a.	b.	c.
Price: $110	Price: $5,000	Price: $90
12% discount	2.4% increase	_____% discount
New price: $_____	New price: $_____	New price: $59

2. The Jefferson family bought three children's tickets and two adult's tickets to the county fair. They got a 5% discount on the total purchase price before tax. Lastly, a 6.2% sales tax was added to the total. If the normal price of a child's ticket is $10 and an adult's ticket is $20, find the cost of tickets for the family.

3. This year the college has 1,210 students—an increase of 6.6% from last year. How many students did the college have last year?

4. Mary's dog weighed 25 kg, but then it got sick and lost 2.3 kg.

   a. What percentage of body weight did the dog lose?

   b. Mary weighs 58 kg. If Mary lost the same percentage of her body weight as what the dog did, how much would Mary weigh?

5. A rectangular playground area measures 5 m by 6.5 m. It is enlarged so that it becomes 7.2 m by 10 m. What is the percentage of increase in its area?

6. In 2010, the United States had 10,779,264 males and 10,320,257 females that were 0 to 4 years old. It also had 10,827,017 males and 11,282,003 females that were 50 to 54 years old.

   a. How many percent more males than females are there in the age group 0-4 years? (Use relative difference/percentage difference.)

   b. How many percent more females than males are there in the age group 50-54 years? (Use relative difference/percentage difference.)

7. A 12″ pizza in Tony's Pizzeria costs $12.99 and in PizzaTown it costs $15.99. How many percent more expensive is the 12″ pizza in PizzaTown than in Tony's Pizzeria?

8. Jacqueline deposited $2,500 into a savings account that pays a yearly interest rate of 4.4%. Calculate how much her account will contain after three years.

9. Michael borrowed $35,000 for ten years. At the end of those years he paid the bank back $65,800. What was the interest rate?

Mixed Review 11

1. Evaluate the expressions. Give your answer as a fraction or mixed number.

a. $\dfrac{3x}{x+7}$, when $x = -4$	**b.** $\dfrac{1-x}{1+x}$, when $x = 7$

2. Eric walks at a constant speed of 2 m/s for eight seconds. Then he runs for the next ten seconds at a constant speed of 5 m/s.

   a. Plot a graph for the distance Eric runs.

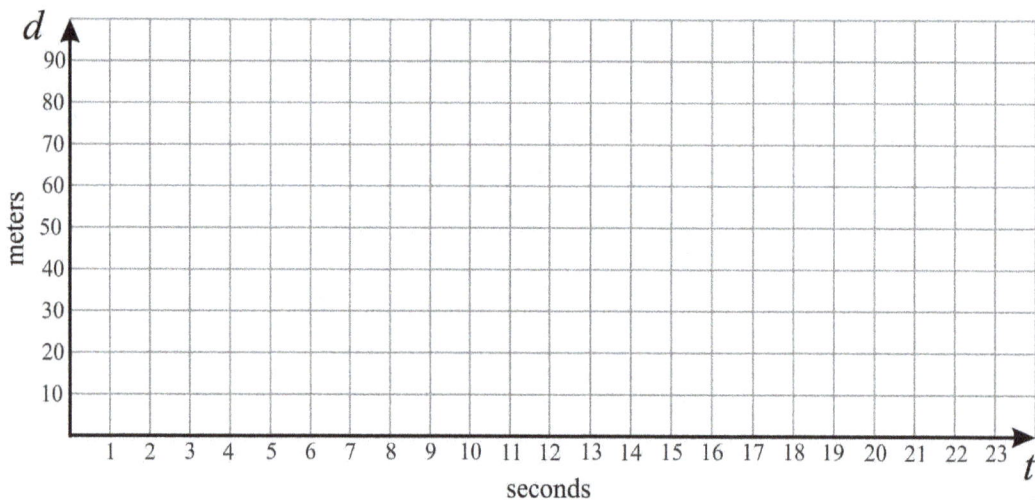

   b. What is the total distance Eric advances?

3. Solve.

a. $\dfrac{6x}{7} = -1.2$	**b.** $6x - 7 = -1.2$

74

4. Are the expressions equal, no matter what values x and y have? If so, you don't need to do anything else. If not, provide a counterexample.

a. $\dfrac{x-y}{3}$ $\dfrac{x}{3} - \dfrac{y}{3}$	b. $x - 2y$ $y - 2x$

5. Evaluate the expression $|a - b|$ for the given values of a and b. Check that the answer you get is the same as if you had used a number line to figure out the distance between the two numbers.

a. a is -5 and b is 6	b. a is -2 and b is -11

6. **a.** Write an expression for the distance between x and 7.

   b. Evaluate your expression for $x = -3$.

7. Write using symbols, and simplify if possible.

   a. the opposite of -2 **b.** the absolute value of -80

   c. the opposite of the sum $6 + 7$ **d.** the absolute value of the sum $-4 + 5$

8. Solve the proportions by using cross-multiplication.

a. $\dfrac{14 \text{ mi}}{0.59 \text{ gal}} = \dfrac{100 \text{ mi}}{V}$	b. $\dfrac{P}{2000 \text{ lb}} = \dfrac{\$4.05}{3 \text{ lb}}$

9. If gasoline costs $3.14 per gallon and if your car gets 21 miles per gallon, find the cost of driving the car for 15 miles.

10. The two figures are similar. Find the length of the unknown side.

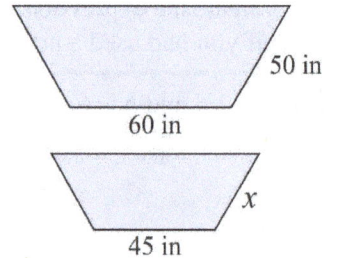

11. Solve using both decimal multiplication and fraction multiplication.

0.6 · 0.7	Decimal multiplication:	Fraction multiplication:

12. Solve using both decimal division and fraction division.

0.24 ÷ 0.5	Decimal division:	Fraction division:

Mixed Review 12

1. Solve the inequalities and plot their solution sets on a number line. Write appropriate multiples of ten under the bolded tick marks (for example, 30, 40, and 50).

a. $\quad 3y + 7 \;<\; 56$	**b.** $\quad -5 + 6z \;\geq\; 175$

2. As a salesperson selling fine art paintings, you are paid a base salary of $180 per week plus $45 per sale. How many paintings do you need to sell in a week in order to earn at least $500?
   Write an inequality for the number of sales you need to make, solve it, and describe the solutions.

3. Matt got a really unreasonable answer for the problem below. Find what went wrong with his solution and correct it .

Jim can swim 30 laps in a pool in 26 minutes.
How many laps could he swim in 45 minutes?

Matt's Answer: He could swim 30 laps.

Solution:

$$\frac{30 \text{ laps}}{26 \text{ min}} = \frac{L}{45 \text{ min}}$$

$$26L = 30 \cdot 26$$

$$26L = 780$$

$$\frac{26L}{26} = \frac{780}{26}$$

$$L = 30$$

4. **a.** Plot the equation $y = 2x - 1/2$.

   b. Plot the equation $y = -(1/3)x$.

   c. Plot the equation $y = 10 - 2x$.

5. Determine whether the point $(-2, 2)$ is on
   the line $y = -(1/2)x + 1$. Show your work.

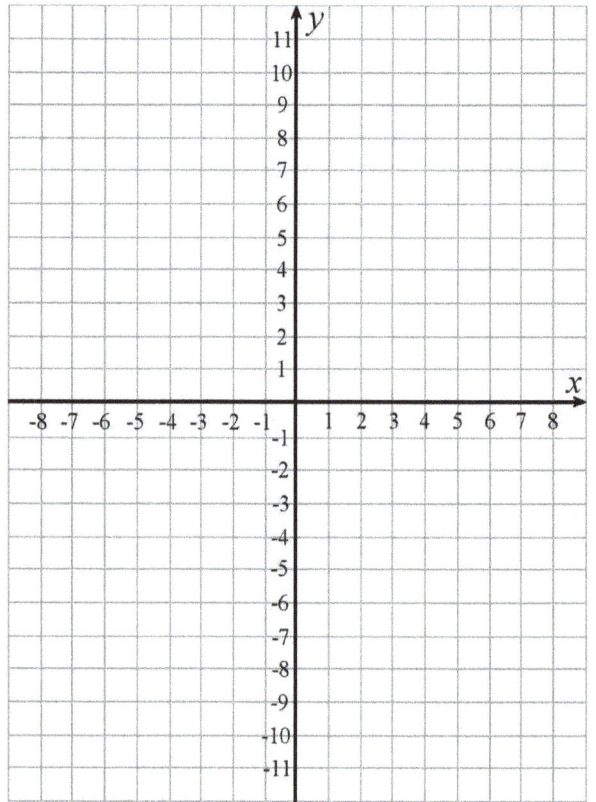

6. The two triangles are similar. How long is the unknown side?

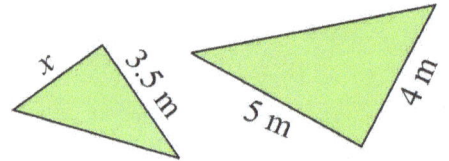

7. **a.** Draw a copy of this rectangle
   using the scale factor 2.5.

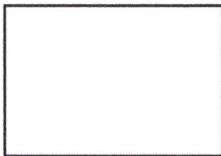

   b. What is the scale *ratio*?

8. **a.** Find the value of the expressions $2p$ and $p - 4$ for different values of p.

Value of p	2p	p − 4	Value of p	2p	p − 4	Value of p	2p	p − 4
−6			−2			2		
−5			−1			3		
−4			0			4		
−3			1			5		

b. Consider the patterns in the values of these expressions.
Is there any value of p for which $2p$ would be equal to $p - 4$?

c. For which values of p in the table is $2p$ more than $p - 4$?

d. For which value of p does the expression $p - 4$ have the value -2?

9. Solve (without a calculator).

a. $10 - \dfrac{5}{6} \cdot 2.7$

b. $0.4 \div \left(\dfrac{2}{9} + \dfrac{1}{3} \right).$

10. Solve.

a. $13 + (-37) + (-8) =$

b. $-94 - (-8) =$

c. $20 - 60 + 90 =$

11. Find the missing numbers.

a. $6 \cdot \underline{\hspace{1cm}} = -42$

b. $-72 \div \underline{\hspace{1cm}} = 8$

c. $\underline{\hspace{1cm}} \div (-12) = -4$

12. Find the value of the expressions when $x = -5$ and $y = 2$.

a. $1 - x^2$

b. $10xy$

c. $-3(x + y)$

Geometry Review

You may use a calculator for every problem in this lesson.

1. The letters from *u* to *z* label the angles in the figure.

   a. Which two angles are complementary?

   b. Which two angles are supplementary?

   c. Which two angles are vertical angles?

   d. If $w = 51°$ and $v = 59°$, then what are the measures of the rest of the angles?

   $u =$ _____ ° $x =$ _____ ° $y =$ _____ ° $z =$ _____ °

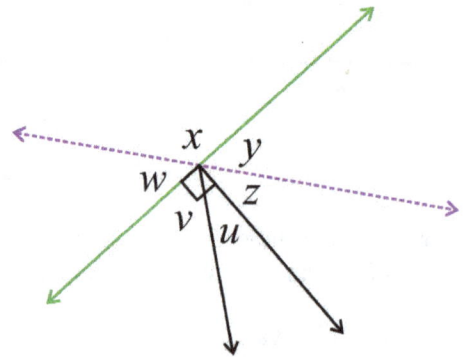

2. In the figure at the right, lines *l* and *m* are parallel.

   a. Write an equation for the measure of the unknown angle *x*.

   b. Solve your equation.

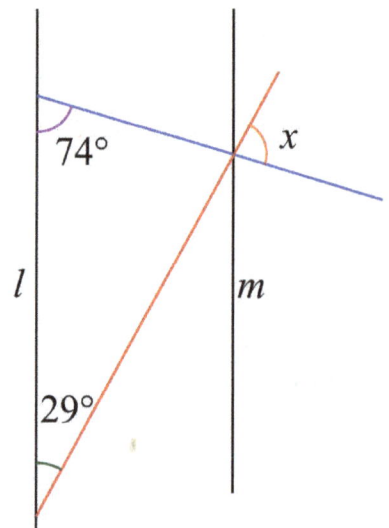

3. The top angle of an isosceles triangle measures 26°.

   a. What do the other two angles measure?

   b. Draw the triangle. Make the base 4 inches long.

4. Draw, using a compass and straightedge only, an isosceles triangle with two sides the length of this line segment:

   _____

5. Draw a triangle with sides 2 in, 2 3/4 in, and 3 3/8 in long.

6. Two sides of a parallelogram measure 10.2 cm and 5.0 cm. There is a 45° angle between them. Do these conditions define a unique parallelogram? If so, draw it. If not, draw several non-congruent (different-shaped) parallelograms that all fit the conditions.

7. This rectangle is a plan for Henry's room drawn at a scale of 1:50. Draw a copy of it at the scale 1:60.

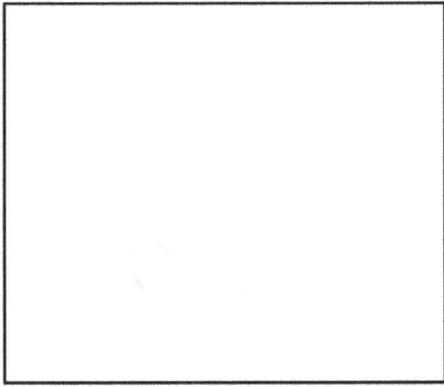

Scale 1:50

8. Draw an altitude into this triangle and then find its area to the nearest square centimeter.

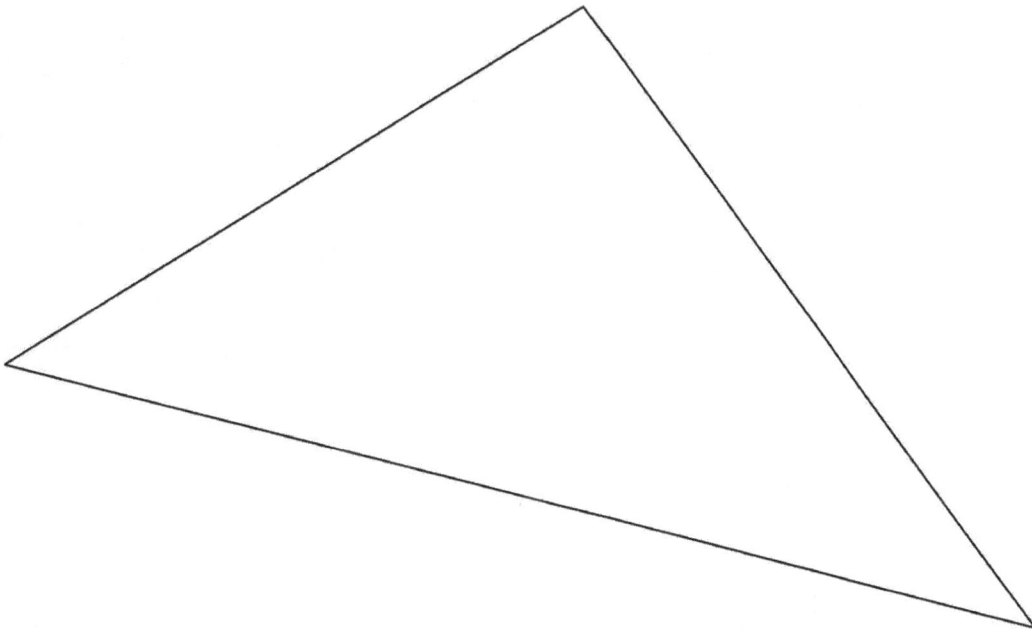

9. **a.** Find the radius of a circle with a circumference of 20.6 cm.

   b. Calculate, to the nearest ten square inches, the area of a circle with a diameter of 8 ft 2 in.

10. Explain in a few words what the pictures at the right are about.

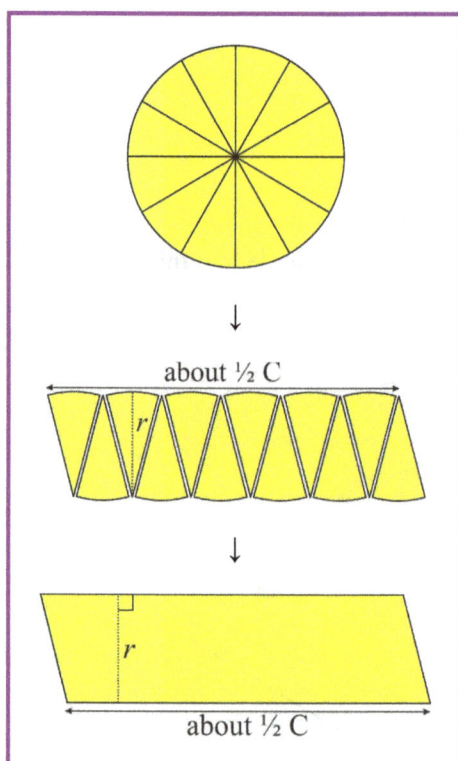

about ½ C

about ½ C

11. The roof of a canopy is in the form of a pyramid. Calculate the total surface area of the roof.

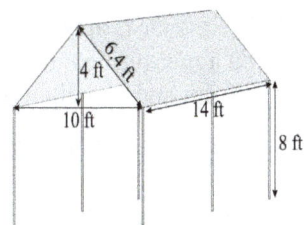

12. Calculate the volume of this box to the nearest hundred cubic inches.

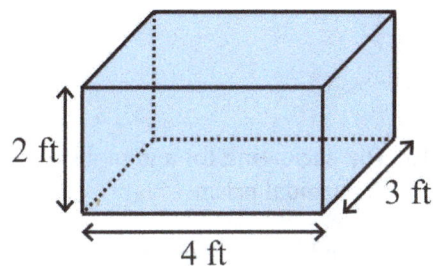

13. **a.** Find the area of this trapezoid in square feet.

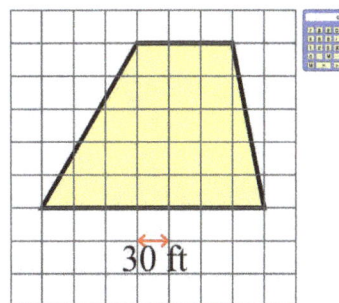

b. Find the area in square yards, too.

14. A glass jar is approximately in the shape of a circular cylinder. Jamie measured its volume using water and a measuring cup, and he found that it held about 530 ml of water. Jamie also measured its diameter as 9 cm and its height as 12 cm.

   a. How many cubic centimeters is 530 ml?

   b. One of Jamie's measurements—either the diameter or height—is in error. How can you tell that?

15. The packaging for a granola bar is in the form of a trapezoidal prism.

   a. Find its volume in cubic millimeters.

   b. Find its volume in cubic centimeters.

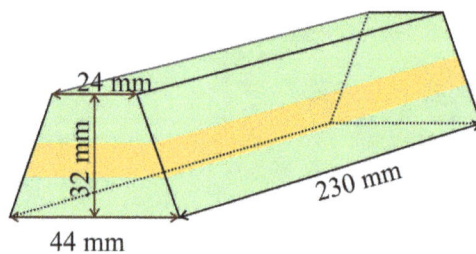

16. What shape is the cross-section?

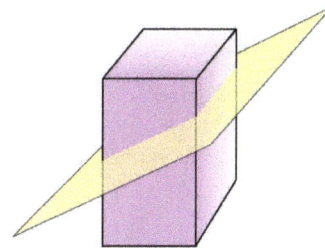

17. Explain or draw a sketch of how to cut through this pyramid with a plane so that the cross-section is a triangle.

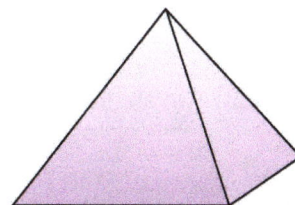

Geometry Test

1. Find the measures of angles *x*, *y*, and *z* without measuring.

   x = _____   *y* = _____   *z* = _____

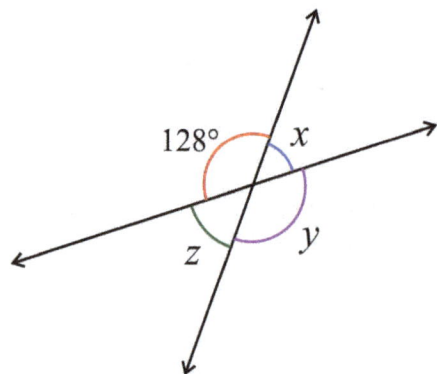

2. Draw two angles that are complementary.

3. Write an equation for the unknown angle. Then solve it. Do not measure any angles.

   Equation for *x*: _____

   Solution:

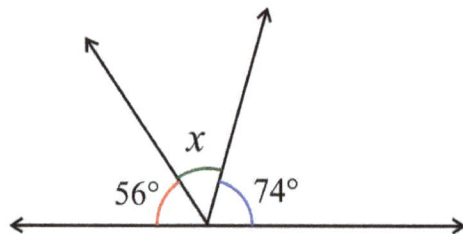

4. Lines *m* and *n* are parallel. Find the measure of angle β without measuring.

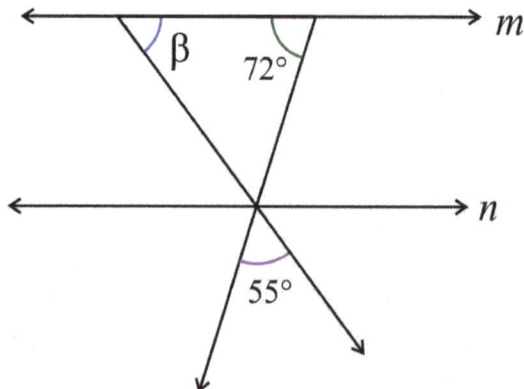

5. An isosceles triangle has an 80° top angle and two 11-cm sides.

    a. Calculate the angle measure of the base angles.

    b. Draw the triangle.

6. Draw two lines that are perpendicular to each other using only a compass and a straightedge.

7. The "Yield" traffic sign is in the shape of an upside-down equilateral triangle. The image below shows the outline of its basic triangular shape, drawn at a scale of 1:10. Redraw it at a scale of 1:12.

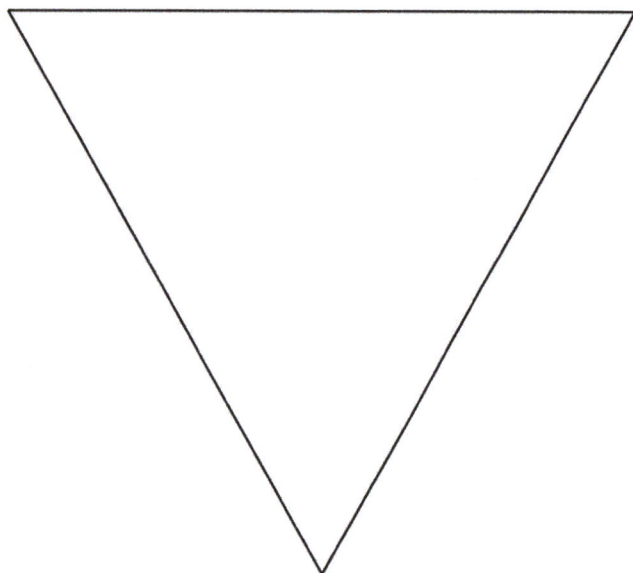

8. One acre is 43,560 square feet. Use that information and your knowledge of geometry to calculate the area of this trapezoidal plot:

   a. in square feet

   b. in acres, to the hundredth of an acre

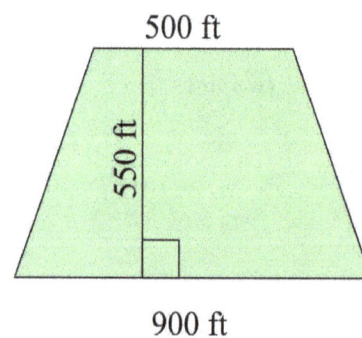

9. A rectangular prism is cut with a plane that is perpendicular to the prism's base. What figure is formed at the cross-section?

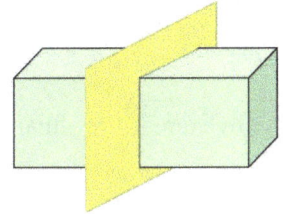

10. A rectangular pyramid is cut with a plane that is parallel to the pyramid's base. What figure is formed at the cross-section?

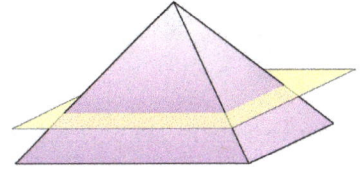

11. A large circular wall clock has a diameter of 40.0 cm. Find its area to the nearest ten square centimeters.

12. Calculate the volume of this water tank in <u>cubic meters</u>.

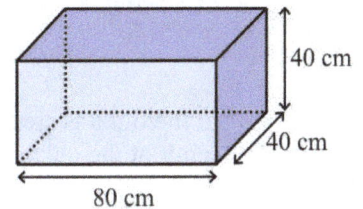

40 cm

40 cm

80 cm

13. Calculate the surface area of this cylindrical bucket for toys to the nearest ten square centimeters.

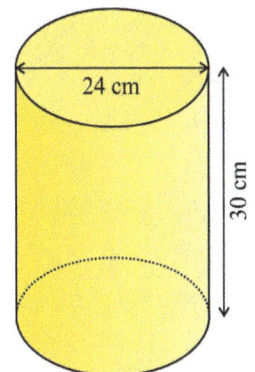

24 cm

30 cm

Mixed Review 13

1. Give a real-life situation for the sum $(-30) + (-12)$.

2. Solve $53 + (-91) + 21 + (-3) + (-55)$.

3. On average, Scott makes a basket nine times out of twelve shots when he is practicing. How many baskets can he expect to make when he tries 200 shots? Use equivalent rates.

$$\frac{9 \text{ baskets}}{12 \text{ shots}} = \frac{\boxed{} \text{ baskets}}{\boxed{} \text{ shots}} = \frac{\boxed{} \text{ baskets}}{200 \text{ shots}}$$

4. The two parallelograms are similar with a similarity ratio of 2:7. Find the length of the unknown side.

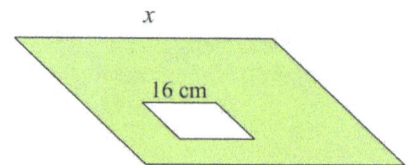

x

16 cm

5. A book costing $15 is discounted by 15%. What is the new price?

6. At 20% off, a tire costs $100. What was its price before the discount?

7. A certain county acquired a total revenue of $135,000 in year 2013. Of it, 41.0% came from property tax and the rest from other sources. In 2014, the total revenue fell by $3,500 and the revenue from property tax by $2,100. What percentage of the county's revenue came from property tax in 2014?

8. Simplify the expressions.

a. $2 + w + 11 + w + 2w$	**b.** $2 \cdot w \cdot 11 \cdot w \cdot 2w$	**c.** $c \cdot c \cdot 3 \cdot d \cdot d \cdot d$

9. Car 1 gets a gas mileage of 20 miles per gallon and car 2 gets 24 miles per gallon.

   a. Write two equations—one for each car—relating the distance (d) driven to the amount of gasoline used. Use g to represent the amount of gasoline.

     Car 1: $d =$ _____   Car 2: $d =$ _____

   b. Plot both equations in the same coordinate grid.

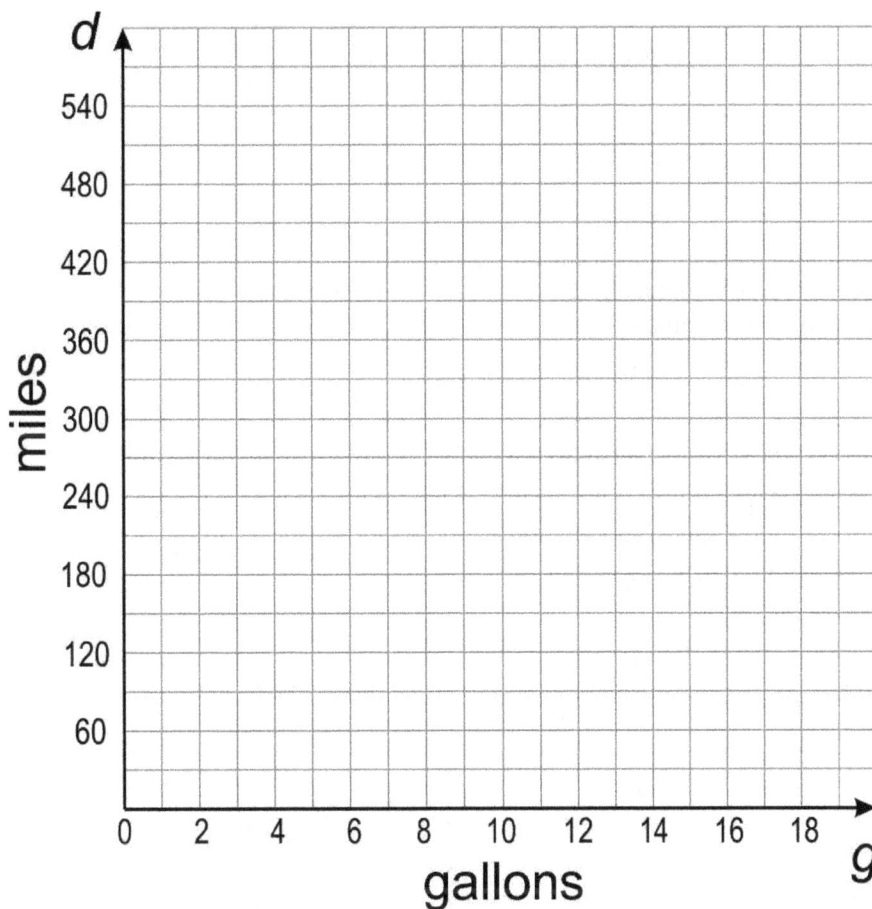

   c. State the slopes of the two lines.

   d. Plot a point on each line that corresponds to the distance 120 miles.

   e. Plot a point on each line that corresponds to the unit rate.

   f. What does the point (0, 0) mean in this situation?

10. To the nearest tenth of a percent, calculate by how many percent these quantities changed.

a. The postage for a letter increased from \$0.78 to \$0.81.	**b.** Jamie sold 445 newspapers last week. This week he sold 487.

11. A rectangle with sides of 2 1/4 in. and 3 in. is enlarged by a scale factor of 3.5. Find the area of the resulting rectangle to the tenth of a square inch.

12. How many 3/4-foot long pieces can you cut out of a roll of string that is 8 3/8 feet long? Write a *quotient* to solve the problem.

$$ \underline{\hspace{2cm}} = $$

13. One of the expressions below gives an area of a rectangle and the other its perimeter.

$$ 14a + 8b \qquad 7a \cdot 4b $$

How long is each side of the rectangle?

14. Solve.

a. $1\dfrac{1}{3} - y = \dfrac{5}{8}$	**b.** $z + \dfrac{2}{3} = 1\dfrac{9}{10}$

Mixed Review 14

1. **a.** Sketch a rectangle with sides $5x$ and $6x$ long.

   b. What is its area?

   c. What is its perimeter?

2. A photo editing software was discounted by 2/5. The discounted price is $29.97.

   a. Find the original price using logical reasoning and/or a bar model.

   b. Choose a variable to represent the original price. Write an equation for the situation and solve it. Compare the solution steps of the equation to the way you solved the problem in (a).

3. Find the percentage of increase or decrease.

a. A flashlight that costs $9 is discounted so that now it costs $8.10. What is the discount percentage?	**b.** A chair used to cost $20, but now it costs $26. What is the percentage of increase?

4. Mason got 16 points out of 21 in a quiz. What is his percentage score, to the nearest tenth of a percent?

5. Mark pays 22.5% of his income in taxes. If he earns $2,350 in a month, find how much he has left after taxes.

6. A farmer gets paid $3 for a bushel of corn. A bushel is about 35.2 liters.

   a. How much does the farmer get for one liter of corn?

   b. Let P be the the amount of money the farmer gets and V be the amount of corn in liters. Write an equation relating the two variables.

   c. Plot your equation. Choose appropriate scaling for the P-axis.
   Hint: calculate how much the farmer gets for 100 liters and for 600 liters of corn.

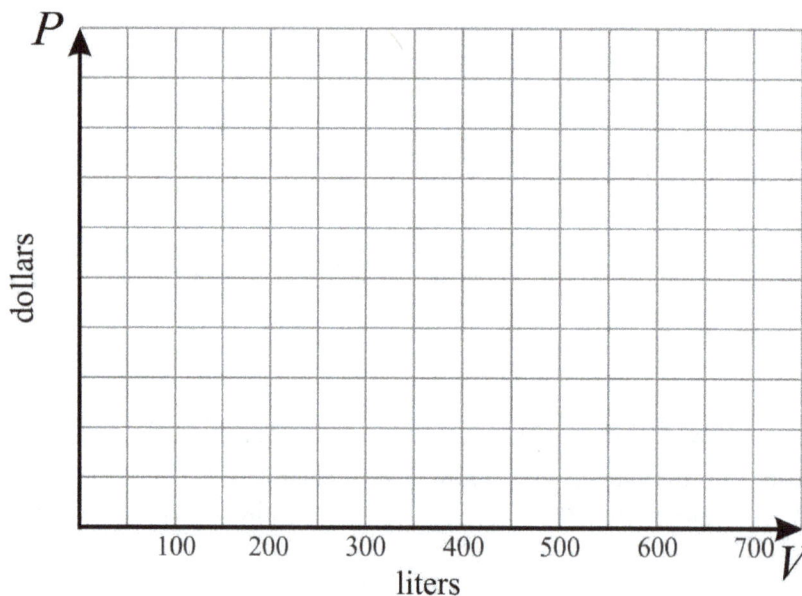

   d. Plot the point corresponding to 200 liters of corn.

   e. Plot the point corresponding to the farmer earning $50.

7. Study the problem below and Jane's solution for it. She figured that exactly half of the books were to be sent to book stores.

A printing press printed 1,500 copies of a book. 5/6 of those were printed as paperbacks and the rest were printed with hard covers. Now, 3/5 of the paperbacks need to be sent to various book stores.	Jane's calculation to solve this:
How many books is that?	$\dfrac{\overset{1}{\cancel{3}}}{\underset{1}{\cancel{5}}} \cdot \dfrac{\overset{1}{\cancel{5}}}{\underset{2}{\cancel{6}}} \cdot 1{,}500 = \dfrac{1}{2} \cdot 1{,}500 = 750$

a. Is her answer correct?

b. Is the way she calculated it correct? If not, correct it.

8. **a.** Draw a rectangle with an aspect ratio of 4:3 (width to height) so that the width is 6 cm.

   b. Then enlarge your rectangle using the scale ratio 3:5. Draw the enlarged rectangle.

   c. What is the aspect ratio of the enlarged rectangle?

Pythagorean Theorem Review

1. Find the square roots.

a. $\sqrt{144}$	b. $-\sqrt{81}$	c. $\sqrt{1{,}600}$
d. $\sqrt{10^2 - 6^2}$	e. $\sqrt{49 \cdot 49}$	f. $\sqrt{5 \cdot (83 - 3)}$

2. **a.** If the side of a square measures $\sqrt{7}$ cm, what is its area?

   b. How long is the side of a square with an area of 20 cm^2?

3. Solve. Give your answer to the nearest thousandth. You may use a calculator.

a. $y^2 + 18 \ = \ 35$	b. $0.6h^2 \ = \ 4$

4. For each set of lengths, determine whether they form a right triangle.

   a. 20, 24, 30

   b. 2.6, 1.0, 2.4

5. Solve for the unknown side of each triangle. Remember, you can ignore the negative answer. (Why?)

a.

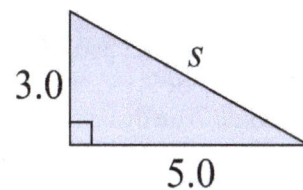

3.0 s 5.0

b.

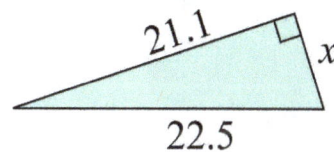

21.1 x 22.5

6. Lauren and Anna want to make this pennant for their jogging club. Calculate its total area.

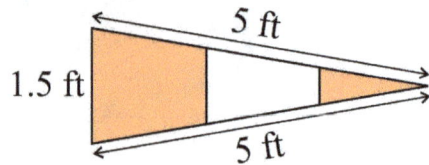

7. The map shows part of downtown Nashville, Tennessee. The triangle ABC on the map is very close to a right triangle. The distance AB is 370 m and the distance AC is 620 m. However, these distances are approximate, so your calculations will also be only approximate.

About how much shorter is it to travel from point A to point C along Lafayette Street than to travel first along Korean Veterans Boulevard and then along 5th Avenue South?

Pythagorean Theorem Test

Do not use a calculator for problems 1-3 of the test.

1. Calculate the values of the square roots.

a. $\sqrt{1{,}000{,}000}$	**b.** $\sqrt{400}$	**c.** $\sqrt{1}$
d. $\sqrt{75-11}$	**e.** $\sqrt{10^4}$	**f.** $\sqrt{53^2}$

2. Do $\sqrt{-9}$ and $-\sqrt{9}$ have the same value? Explain.

3. Solve the equations. Round the answers to three decimals.

a. $\quad s^2 - 17 \;=\; 19$	**b.** $\quad 5y^2 \;=\; 89 + 36$

4. Determine whether the lengths 13.4 m, 7 m, and 10.2 m form a right triangle using the Pythagorean Theorem.

5. Find the length of the unknown side.

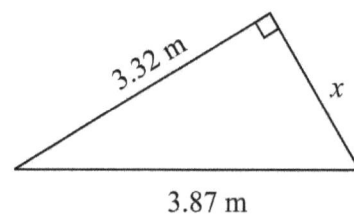

3.32 m

x

3.87 m

6. Find the length of a diagonal of a square with 50-cm sides.

7. Calculate the area of this shape to the nearest tenth of a square foot.

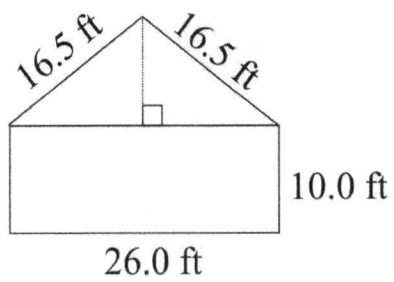

16.5 ft   16.5 ft

10.0 ft

26.0 ft

Mixed Review 15

1. A farmer plowed three acres of his field, which is 2/5 of the total area he is planting this year. How many acres is he planting this year?

   a. Choose a variable for the unknown quantity, and write an equation for the problem.

   b. Solve the equation.

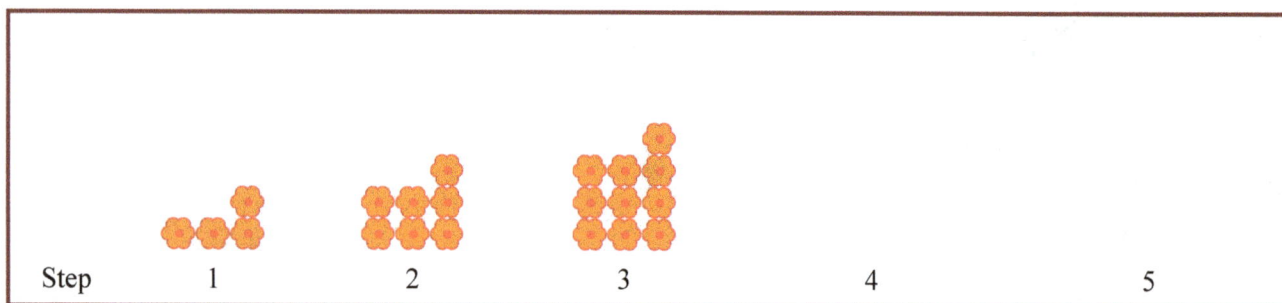

| Step | 1 | 2 | 3 | 4 | 5 |

2. **a.** Draw the next two steps.

   b. How do you see this pattern growing?

   c. How many flowers will there be in step 39?

   d. What about in step n?

3. Fill in the missing parts in this justification for the rule *"Negative times negative makes positive."*

   (1) Substitute $a = -1$, $b = 1$, and $c = -1$ in the distributive property $a(b + c) = ab + ac$.

   _____ (_____ + _____) = _____ · _____ + _____ · _____

   (2) The whole left side is zero because _____ + _____ = 0.

   (3) So the right side must equal zero as well.

   (4) On the right side, $-1 \cdot 1$ equals _____. Therefore, $-1 \cdot (-1)$ must equal _____ so that the sum on the whole right side will equal zero.

4. An educational website currently offers a subscription plan with a $14.95 monthly cost. The company is considering increasing it to $19.95. What is the percent of increase?

5. **a.** Write an equation for angle x and solve it.

   b. Write an equation for angle y and solve it.

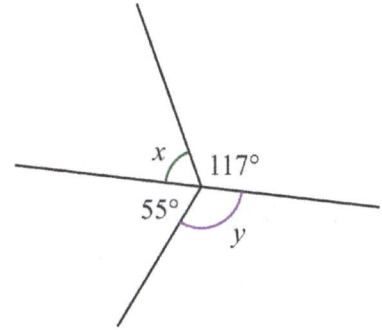

6. Calculate the area of a rectangle with sides 2 yards and 5 yards both in square yards and in square feet.

   A = _____ sq. yd. = _____ sq. ft.

7. Jane has a number of identical small cylinder-shaped water glasses. Their bottom diameter measures 6 cm, and their height is 8 cm. Jane fills them 3/4 full. How many glasses can Jane fill out of a 1-liter pitcher of juice? *Hint: 1 cm³ = 1 ml.*

8. **a.** Draw a parallelogram that has 36° and 144° angles and sides that are 7 cm and 4.5 cm long.

   b. Find the area of the parallelogram.

9. A farmer gets paid $3 for a bushel of corn. A bushel is about 35.2 liters. How much would the farmer be paid for 200 liters of corn?

10. Angela and Mary solved the following problems differently.

   a. Who got each question correct?

   b. Which package of granola costs less by weight?

A package of granola weighs 800 g and costs $3.60. Another package weighs 600 g and costs $3.00.
(1) How many percent heavier is the first than the second?
(2) How many percent more expensive is the first than the second?

Angela: (1) I subtract 800 g − 600 g = 200 g and write the fraction $\dfrac{200}{800} = \dfrac{1}{4} = 25\%$.

Angela: (2) I subtract $3.60 − $3.00 = $0.60 and write the fraction $\dfrac{0.60}{3.60} = \dfrac{1}{6} \approx 17\%$.

Mary: (1) I subtract 800 g − 600 g = 200 g and write the fraction $\dfrac{200}{600} = \dfrac{1}{3} \approx 33\%$.

Mary: (2) I subtract $3.60 − $3.00 = $0.60 and write the fraction $\dfrac{0.60}{3.00} = \dfrac{1}{5} = 20\%$.

11. Calculate.

a. $8\dfrac{3}{4} \div 2\dfrac{5}{8} + 2\dfrac{2}{5}$	**b.** $4\dfrac{2}{7} \div \dfrac{6}{7} \cdot \dfrac{5}{8}$

Mixed Review 16

1. **a.** In the following problem, represent Jayden's salary with a variable and write an equation for the situation.

Jayden pays 1/7 of his salary in taxes. If he paid $415 in taxes, how much was his salary?

   b. Solve the equation.

2. Factor these expressions (write them as products).

a. $7x + 21 \ = \ $ ___ (___ + ___)	**b.** $24w - 16 \ = \ $ ___(___ − ___)
c. $-21t - 7 \ = \ -7($ ___ + ___)	**d.** $50a - 70b - 120 =$
e. $-55a + 30 \ =$	**f.** $-56y - 84 - 7x =$

3. Ava used 3 1/2 cans of paint to paint 2/3 of a room.

   a. Write the unit rate as a complex fraction and simplify it.

   b. Using paint at the same rate, how much more paint does Ava need to finish painting the room?

4. Evaluate the expressions.

a. $100 - x^2$, when $x = -2$	**b.** $\dfrac{2w}{w+3}$, when $w = 1/2$

5. Solve.

a. $x - \dfrac{5}{6} = 7\dfrac{1}{3}$	**b.** $2\dfrac{1}{4} - w = 1\dfrac{2}{7}$
c. $5y = -\dfrac{4}{9}$	**d.** $v + \dfrac{1}{5} = -\dfrac{1}{12}$

6. The price of a standing lamp increased from $19 to $22.50. What was the percent of increase?

7. Mason took a $1,500 loan at 9.8% annual interest rate to purchase a computer. He paid it back 1 1/2 years later. Calculate the total amount Mason paid at that point.

8. The table below shows how long it takes for a car to travel a distance of 120 km at different speeds.

Speed (km/h)	120	100	80	60	40	20
Time (h)	1	1.2	1.5	2	3	6

    a. Are these two quantities, speed and time, in proportion?

      Explain how you can tell that.

    b. If so, write an equation relating the two and state the constant of proportionality.

9. The four children in the Adams family have earned the following points in a computer game.

Child	Points
Chris	365
Grace	458
Hailey	602
Tony	553

    a. Comparing Grace and Hailey, how much better percentage-wise is Hailey doing than Grace? Use relative difference.

    b. Comparing Tony and Chris, how much better percentage-wise is Tony doing than Chris? Use relative difference.

10. **a.** Draw an equilateral triangle using only a compass and a straightedge. Make the side length whatever you wish. If you draw a small one, you can draw it here. Or, you can use blank paper.

    b. Draw the altitude into it.

    c. Use a centimeter ruler to measure what you need, and find its area to the nearest square centimeter.

11. A triangle has 45° and 60° angles and a 5.6-cm side between those angles. Does the information given define a unique triangle? If it does, write yes, and draw the triangle.

If not, prove that it doesn't by sketching at least two non-congruent (different-shaped) triangles that satisfy the given conditions.

12. Three mirrors are set up in a form of an equilateral triangular prism. They will go inside a tube to make a kaleidoscope. Calculate the volume of the triangular prism. Round to the nearest ten cubic centimeters.

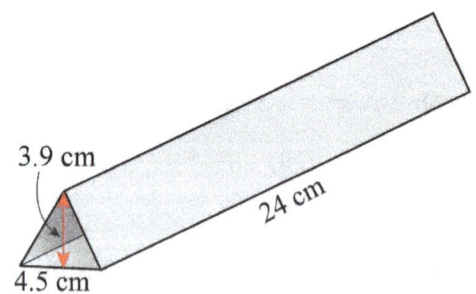

3.9 cm

24 cm

4.5 cm

Probability Review

1. The chart lists the favorite school subjects of the students in a 7th grade classroom.

   a. You choose one student randomly. What is the probability that the student's favorite subject is not math, English, or science?

   b. What is the probability that a randomly chosen student's favorite subject is math?

   c. Now look at the boys only. If you choose one boy randomly, what is the probability that his favorite subject is math?

   d. If you choose a girl randomly from among the girls, then what is the probability that her favorite subject is math?

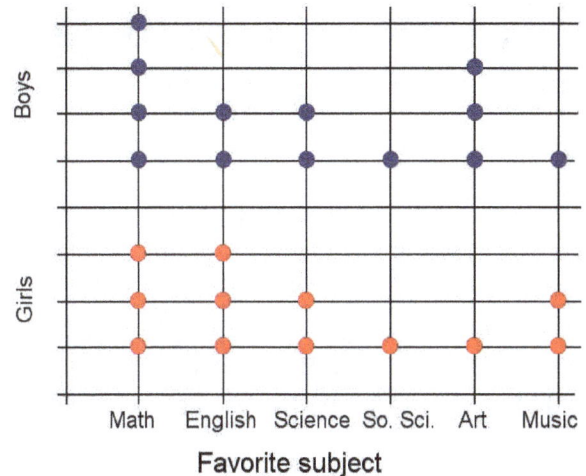

2. Abigail rolls two dice. Find the probabilities of these events:

   a. P(5, 6)

   b. P(even, even)

   c. P(at least 5, at least 5)

   d. P(at most 3, at most 2)

3. Julie and Jane experimented with rolling a die. They rolled the die 60 times in a row and recorded the results:

   2 4 3 3 5   2 5 4 1 6   1 4 4 4 5   4 2 6 4 1   5 2 1 1 1   2 1 3 6 2
   4 2 2 3 3   5 2 3 6 4   4 4 1 3 2   6 6 4 5 6   4 6 5 6 5   6 5 6 5 3

   a. In this experiment, what was the probability of rolling a 1?

   Rolling a 4?

   b. Why are those probabilities not 1/6?

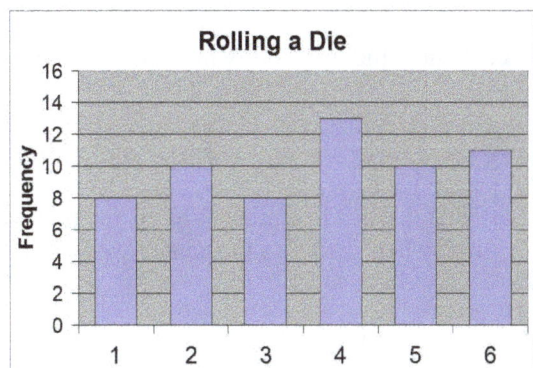

4. Andrew did an experiment where he tossed two coins 200 times and recorded the outcomes. The table below shows his results. "H" means "heads," and "T" means "tails," so "HH," for example, means both coins landed "heads."

   a. Calculate the experimental and theoretical probabilities and fill in the table.

Outcome	Frequency	Experimental Probability (%)	Theoretical Probability (%)
HH	38		
HT	53		
TH	46		
TT	63		
TOTALS	**200**		

   b. How would the experimental probabilities change if Andrew redid this experiment with 2,000 tosses?

5. In a multi-player computer game, a computer chooses colors randomly for a girl's dress. The first color it chooses is the main color of the dress. The second color is for the bows and some layers of the skirt. The computer uses this list of colors: *red, blue, purple, pink, orange, yellow, mint.*

   After choosing the main color, the computer removes it from the list and chooses the second color from the resulting list of six colors. That way the dress is sure to have two different colors.

   a. What is the probability the computer chooses first purple, then orange?

   b. What is the probability the computer chooses first red, then not pink?

   c. Janet doesn't like mint. What is the probability her character gets a dress with no mint in it when she plays the game?

Probability Test

You may use a calculator for all the problems in this test.

1. You roll a number cube with numbers 1, 2, 3, 4, 5, and 6 printed on the faces.
   Find the probabilities as fractions.

   a. P(not 5)

   b. P(2 or 6)

   c. P(less than 9)

   d. P(not 2 nor 5)

2. Two spinners are spun.

   a. In the space below, draw a tree diagram showing all the
   possible outcomes of this experiment.

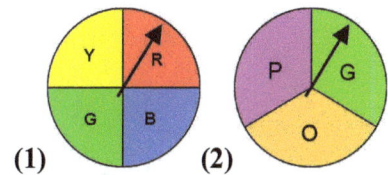

(1)   (2)

Then find the probabilities.

b. P(yellow; purple)

c. P(red or yellow; orange)

d. P(not red; not orange)

3. Two dice are rolled. Find the probabilities of these events:

   a. You get a sum of six on the two dice.

   b. You get less than 3 on each dice.

   c. One dice is 6 and the other is not (in either order).

4. Logan and Alex tossed two coins 400 times.

   a. List all the possible outcomes when two coins are tossed just one time.

   b. Here are Logan's and Alex's results. Calculate and fill in the table the experimental and theoretical probabilities to the nearest tenth of a percent.

	Frequency	Experimental probability	Theoretical probability
TT	5		
TH	8		
HT	182		
HH	205		
TOTALS	400		

   c. Suggest a reason for the large discrepancy between the experimental and theoretical probabilities.

5. What is the probability of getting tails, tails, tails when you toss a coin three times in a row?

6. Lily and Grace placed some stuffed animals in a bag. Then they randomly pulled out one animal and put it back, and repeated this 120 times. Here are their results:

Animal	Frequency
Elephant	58
Giraffe	29
Bear	17
Cat	11
Bird	5
Totals	**120**

a. Based on their results, what is the approximate probability of pulling the cat out of the bag?

b. If this experiment was repeated 300 times, approximately how many times should they expect to get the bear?

Mixed Review 17

1. Beth got a really unreasonable answer. Find what went wrong with her solution and correct it.

Eighty liters of blueberries costs $35.
How much would 52 liters cost?

Beth's Answer: 52 liters would cost $118.86.

Beth's Solution:

$$\frac{80}{35} = \frac{C}{52}$$

$$35C = 80 \cdot 52$$

$$35C = 4160$$

$$\frac{35C}{35} = \frac{4160}{35}$$

$$C = 118.86$$

2. Write an equation using a variable for the unknown angle, and solve it.

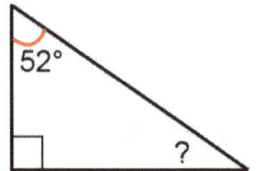

52°

?

3. Determine whether the lengths 11.4 cm, 19 cm, and 15.2 cm form a right triangle.

4. Calculate the volume of this box both in cubic centimeters and in cubic meters.

V = _____ cm^3

V = _____ m^3

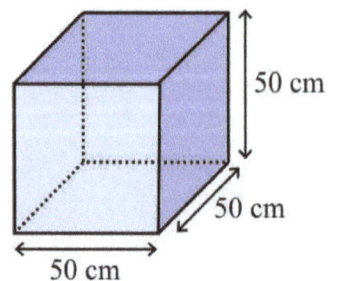

50 cm

50 cm

50 cm

116

5. **a.** What kind of triangle is this?

   b. Find its area to the nearest ten square centimeters.

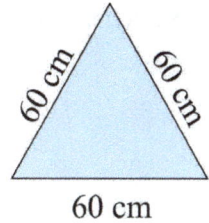

6. Margie drew this circle on a sheet of paper that measures 21 cm by 29.7 cm.

   a. Find the area of the circle to the nearest square centimeter.

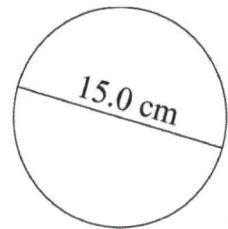

   b. What percentage of the area of the paper does the circle occupy? Give your answer to the nearest tenth of a percent.

   Note: don't use the rounded result from part (a) for this question as that may throw off your answer. For this calculation, you need to keep several decimal digits for the area of the circle.

7. A rectangle with sides of 3 1/2 in. and 2 in. is enlarged in a ratio of 2:5.

   a. Find the lengths of the sides of the resulting larger rectangle.

   b. Find the area of the resulting rectangle.

8. Use the distributive property to multiply.

a. $7(x + 8) =$	**b.** $4(2y - 10) =$	**c.** $0.1(2x + 18) =$

9. Sketch a rectangle with an area of $5x + 15$ and label the lengths of its sides.

10. Solve.

a. $\dfrac{v - 6}{7} = -31$	**b.** $\dfrac{x}{4} - 1 = -5$

11. In May, a bookstore sold 2,400 books. In June it sold 2,000 books. By what percentage did the book sales decrease?

12. The students in Seventh Grade took part in a math contest. Henry scored 77 points, and Mary got 86 points. The average score of all the students in the contest was 66 points.

    a. How much better (in percent) was Henry's score compared to the average score?
    Hint: Use relative difference.

    b. How much better (in percent) was Mary's score compared to the average score?

    c. How much better (in percent) did Mary score than Henry?

13. Solve. If the result is a fraction, simplify it to lowest terms.

a. $3 \cdot (-3) \cdot (-1) =$	**b.** $(-6) \cdot 8 \div 16 =$	**c.** $-8 \cdot (-2) \cdot (-5) =$
d. $(-42) \div (-7) =$	**e.** $7 \div (-42) =$	**f.** $-8 \div (-2) + (-5) =$

Mixed Review 18

1. Find the square roots (without a calculator).

a. $\sqrt{1}$	b. $\sqrt{64}$	c. $\sqrt{10,000}$	d. $\sqrt{400}$

2. The floors of a skyscraper are 11 feet apart. The bottom floor, which is actually the basement, is located 9 feet below the ground.

   a. Write an expression that tells you the elevation of the *n*th floor.

   b. At which elevation is the 47th floor?

   c. Write an equation to find which floor is at an elevation of 288 feet, and solve it.

3. Explain how we can use the pictures at the right to show that the area of a circle is $A = \frac{1}{2}C \cdot r$, where C is the circumference and *r* is the radius of the circle.

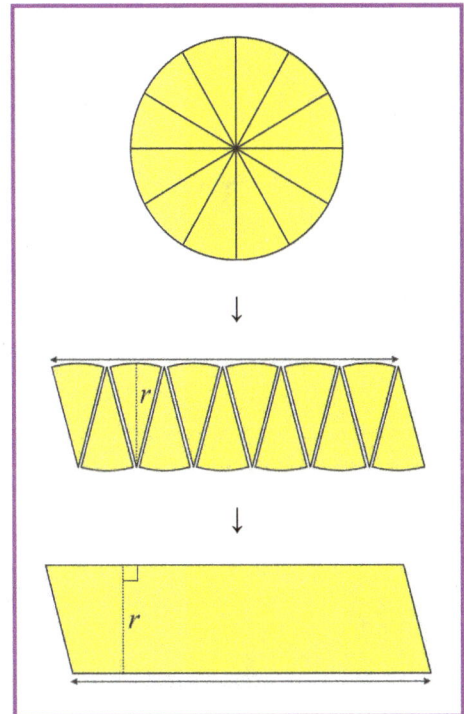

4. In an educational game, you earn 3 play coins for each 10 problems you complete.

    a. Give the unit rate as problems per coin.

    b. How many problems would you need to solve in order to earn 75 coins so you can purchase a phone for your virtual character in the game?

    c. In the same game, if you have solved 376 problems, how many coins have you earned?

5. Which covers more of the wall, a square clock with 30-centimeter sides or a circular clock with 36-centimeter diameter?   How much more?

6. Alison pays \$14.95 per month for an Internet service and she can use 200 gigabytes of bandwidth in that time. For any bandwidth she uses in excess of 200 gigabytes, she will pay \$0.35 per gigabyte. She used 75% of her monthly bandwidth quota in 18 days in September. If she continues to use bandwidth at the same rate, how much extra will she pay at the end of the month?

7. The two figures are similar. Calculate the length of side marked with x.

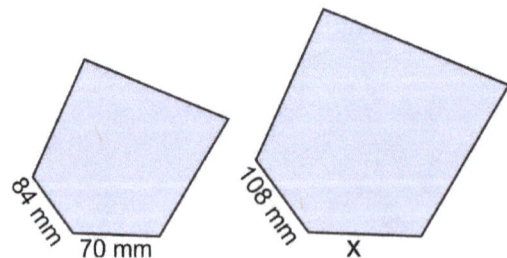

84 mm   70 mm   108 mm   X

8. Solve for the unknown side of the triangle to the nearest tenth of a foot.

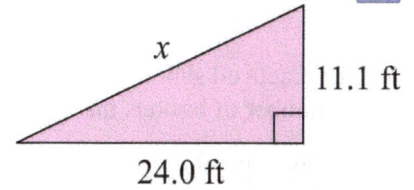

x

11.1 ft

24.0 ft

9. A hexagonal prism is 12 cm high. Its base is a regular hexagon with a side of 2.5 cm and an area of 16.2 cm^2.

   a. Calculate the surface area of the prism.
   Hint: Sketch the prism to help you.

   b. Calculate the volume of the prism.

10. You choose the digits for a two-digit number randomly from the digits 2, 3, 5, and 7 (prime numbers). Each digit can be used twice; for example, it is possible to make 55.

   a. What is the probability of making a number between 31 and 40?

   b. What is the probability of making a number where both digits are the same?

Statistics Review

1. Jake practiced shooting baskets on ten different days. In each practice he shot 50 times, and he recorded the number of baskets that he made. Sadly, he felt he didn't improve any.

Mon	Tue	Wed	Thu	Fri	Sun	Mon	Tue	Thu	Fri
26	33	30	34	29	27	33	27	31	35

   a. Draw a boxplot of the data.

   b. Based on this data, estimate how many baskets Jake would make in 120 shots.

2. Harry belongs to a club that helps abandoned animals. Harry would like for his club to purchase some new equipment, and he wants to find out whether the other members of the club support his idea. Harry plans to interview those club members who stay a little longer after their regular meeting to chat.

   a. Explain why Harry's sampling method is biased.

   b. Suggest an unbiased random sampling method for him.

3. The algebra teacher, Mrs. Riley, had hundreds of test papers to grade and she knew she couldn't finish grading them in one evening. She decided to take a sample of 20 papers from 7th grade students and another sample of 20 papers from 8th grade students to get an idea of how well the students did.

These are the test results for the two samples:

GRADE 7: 56 59 61 64 66 68 68 73 75 75 76 77 79 79 83 84 88 90 96 97

GRADE 8: 52 54 55 58 60 62 62 63 65 68 69 70 72 72 74 77 82 85 92 99

a. Make a back-to-back stem-and-leaf plot of the results.

b. Just looking at the two distributions, does either grade appear to have done better on the test? If so, which one?

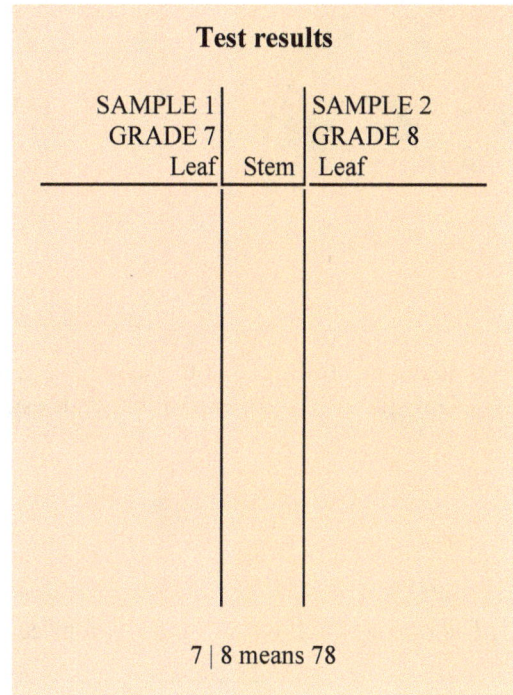

Does either grade appear to have a greater variability in its test results? If so, which one?

Test results

SAMPLE 1 GRADE 7		SAMPLE 2 GRADE 8
Leaf	Stem	Leaf

7 | 8 means 78

c. Find the range, median, and the interquartile range for each sample.

Grade 7:   Median _____   Range _____   Interquartile range: _____

Grade 8:   Median _____   Range _____   Interquartile range: _____

d. Do these values support your answers in (b)?

4. The table below shows the results of two separate surveys where students were asked who they would vote for in an upcoming school election.

	Hanley	Johnson	Garcia	Wilson	Evans	Totals
Survey 1	5	25	22	13	10	75
Survey 2	3	23	24	16	9	75

a. What can you infer based on these results?

b. Estimate how many of the school's 1,230 students will vote for Wilson.
How far off do you expect that your estimate might be?

5. The boxplots below have to do with prices of eyeglasses in two different stores.

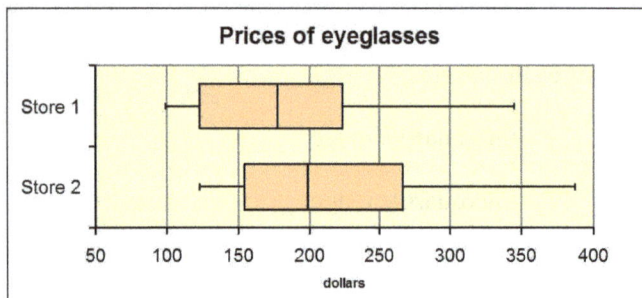

a. Based on the boxplots, does either store appear to have cheaper glasses? If so, which one?

Does either store appear to have greater variability in its prices? If so, which one?

b. Is the difference in the prices significant?

Justify your reasoning.

Statistics Test

1. Researchers compared two different methods for losing weight by assigning 50 overweight people to use each method. The side-by-side boxplots show how many pounds people in each group lost.

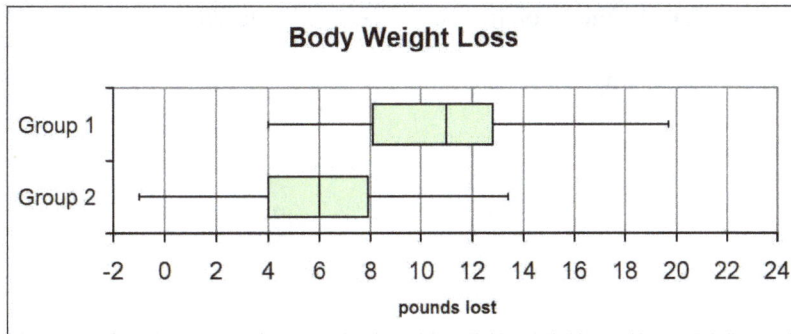

a. Just looking at the two distributions, which group, if any, appears to have lost more weight?

b. Which group, if any, appears to have a greater variability in the amount of weight lost?

c. In group 2, there is one person whose weight loss was −1 pound. What does that mean?

d. Is one of the weight loss methods significantly better than the other?

If so, which one?

Justify your reasoning.

2. Erica is studying the favorite hobbies of adults who live in a small town. She needs a sample of 120 people for her study. Below are listed four possible sampling methods that she could use. <u>Three</u> of the four methods would likely produce a biased sample. Explain which ones they are and why each method is biased.

Sampling method	Biased or not?
(1) Erica generates 120 random numbers between 1 and 1,000, and chooses the corresponding people that she meets on the main street of the town. If method (1) is biased, explain why:	
(2) Erica chooses randomly 120 people from a list of the town's residents. If method (2) is biased, explain why:	
(3) Erica places a box and her survey papers in the town's library, and anyone who wants to can fill in the survey questionnaire. If method (3) is biased, explain why:	
(4) Erica chooses randomly 120 people from a list of the town's taxpayers. If method (4) is biased, explain why:	

3. An ice cream shop surveyed its customers to find out which new flavors of ice cream to add to their selection. They obtained two samples using their customer database. Here are the results:

	Pineapple	Cappuccino	Peanut Butter	Kiwi	Blueberry	Totals
Sample 1	12	32	15	8	13	80
Sample 2	15	36	10	4	15	80

What can you infer from the results?

4. The two histograms show the age distributions of two groups of people. Below you find the means and the mean absolute deviations for both groups.

a. How much do the means differ?

b. Is the difference in the mean ages significant?

Justify your reasoning.

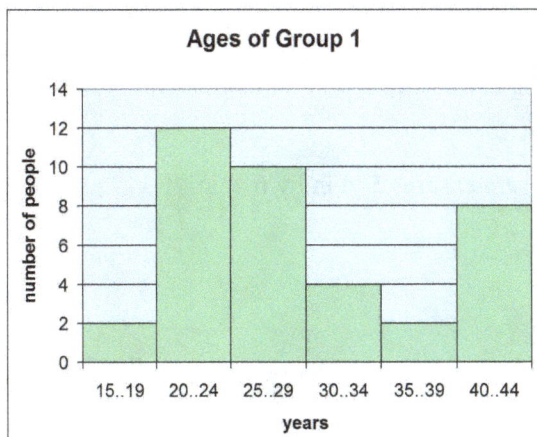

mean 28.6 years   MAD 6.78 years

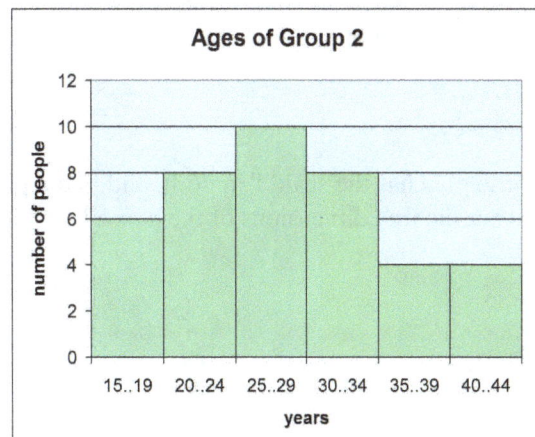

mean 27.55 years   MAD 6.205 years

Mixed Review 19

1. Circle the equation that matches the situation.

   The price of a pair of rubber boots is discounted by 1/5, and now they cost $9.40.

$p - 1/5 = \$9.40$	$\dfrac{p}{4} = 5 \cdot \$9.40$	$\dfrac{4p}{5} = \$9.40$	$\dfrac{p}{5} = \$9.40$	$\dfrac{5p}{4} = \$9.40$

2. Solve the equation in question (1).

3. A juicer that costs $200 is discounted by 15%. Then it is discounted by 20% off of the already lowered price. Find the discount percentage if the price had been discounted from $200 to the final price in one single decrease. Note: The answer will *not* be 35%.

4. A cylindrical can of sardines has a bottom diameter of 6.6 cm and height of 8.5 cm. Another can has a diameter of 10 cm and height of 5.8 cm.

   a. Calculate the volumes of both cans to the nearest cubic centimeter.

   b. How many percent bigger is the larger can than the smaller one?

5. A house plan has the scale 1 in : 6 ft, and in the plan the house measures 5 ¼ in by 6 ¾ in. What are the true dimensions of the house?

6. The picture shows two pairs of intersecting parallel lines.

   a. In the picture, mark with a single arc all the angles that are equal to angle A.

   b. Mark with a double arc all the angles that are equal to angle C.

   c. If angle A is 102°, how many degrees is angle C? _____°

   d. What is the name of the quadrilateral that is enclosed by the two pairs of parallel lines?

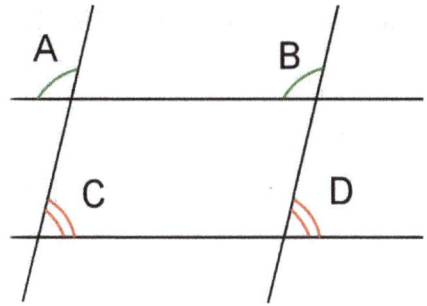

7. An animal mini-puzzle measures 8″ × 5 3/4″ when finished. How many of those puzzles can fit on a 4 ft by 4 ft table?

8. This is a scale drawing of a toy boat, drawn at a scale of 1:8. Redraw it at a scale of 1:5.

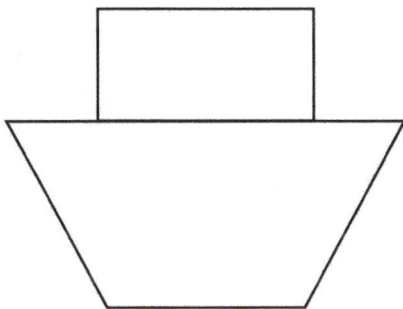

9. **a.** Draw a triangle with sides 2 inches, 3 1/4 inches, and 3 3/4 inches long.

b. Draw a triangle with sides 7 cm, 3 cm, and 3.5 cm long.

10. A spinner with four colors is spun twice.

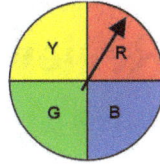

   a. In the space on the right, make a table, a list, or a tree diagram showing all the possible outcomes of this experiment.

   Then find the probabilities:

   b. P(blue; blue)

   c. P(green; not green)

   d. P(not blue; yellow)

   e. P(yellow or green; red or blue)

   Sample space:

11. Tara tosses two coins.

   a. Give an event in this experiment that has the probability of zero.

   b. Give an event in this experiment that has the probability of 1/2.

12. John and Jim decided to check if a particular die was fair or not (in other words, if perhaps it was weighted on one side). They rolled that die 1,000 times. Their results are at the right.

   Based on the results, John said, "Yes, this die is indeed weighted, because we rolled '1' many more times than we rolled '6'."

   Is his conclusion correct? Why or why not?

Outcome	Frequency
1	178
2	160
3	167
4	175
5	167
6	153

13. Sam is studying how well the people in his city like the paintings of the Romantic era. He is planning to stand on a certain street corner near his home and ask passersby if they would like to take part in his study. Explain why his sampling method is biased.

Mixed Review 20

You may use a basic calculator for all the problems in this lesson.

1. The line graph shows the number of candles that a candle factory sold in years 2010 to 2015. Note that the scale is given in "thousand candles."

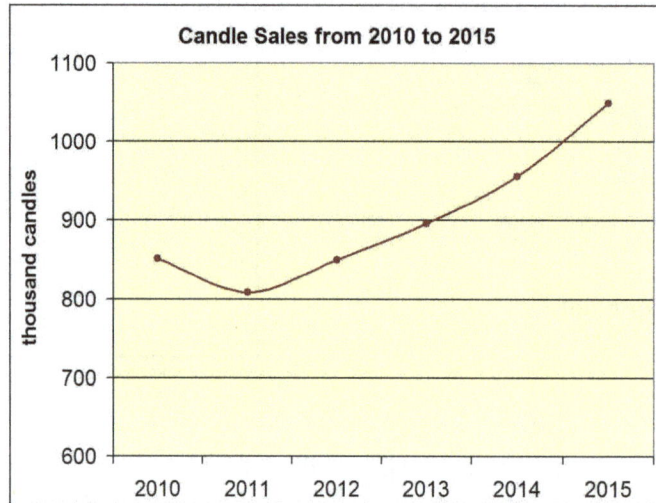

Candle Sales from 2010 to 2015

Estimating the amounts from the graph, calculate the approximate percentage increase in the candle sales from 2010 to 2015.

2. The shape on the left was scaled to become the shape on the right. Find the scale ratio, and then write it as a scale factor. Use a ruler that measures in centimeters.

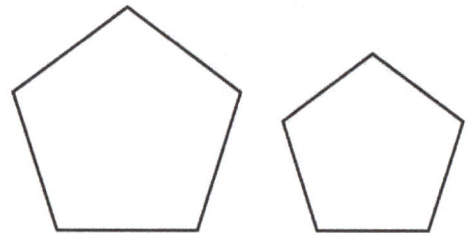

3. **a.** Are these two typing rates equal: 60 words per 90 seconds and 135 words per 3 minutes?

   b. If not, calculate how many more words in 5 minutes would you type at the faster rate than at the slower rate.

4. A family paid $40 for a meal in a restaurant. They gave the waiter a 5% tip (on the $40). They also paid a sales tax of 6.8% (on the $40) included in the total cost. What was the total cost they paid?

5. A ticket to ride a roller coaster costs $5 and a ticket for the bumper cars costs $4.50.

   a. How many percent more expensive is the first ticket than the second?

   b. How many percent cheaper is the second ticket than the first?

6. If the edges of this cube measure 2 inches, how many such cubes do you need in order to have 1 cubic foot?

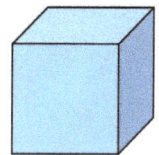

7. Solve the equations.

a. $\dfrac{x}{5} = -4.08$	**b.** $\dfrac{w}{-0.2} = -0.4$
c. $2x + 7 = -4(x + 5)$	**d.** $\dfrac{x + 1}{4} = -2$

8. The principal of a school wants to ask the students' parents whether some extra money should be used to purchase more books for the library, upgrade the computer systems, or to improve the sports facilities of the school. He randomly surveyed some parents who were attending a basketball game at the school.

   Is the principal's sampling method biased or unbiased?

   Explain why.

9. The dot plots show the amount of sugar in fruit and pop drinks (250 ml portion of drink).

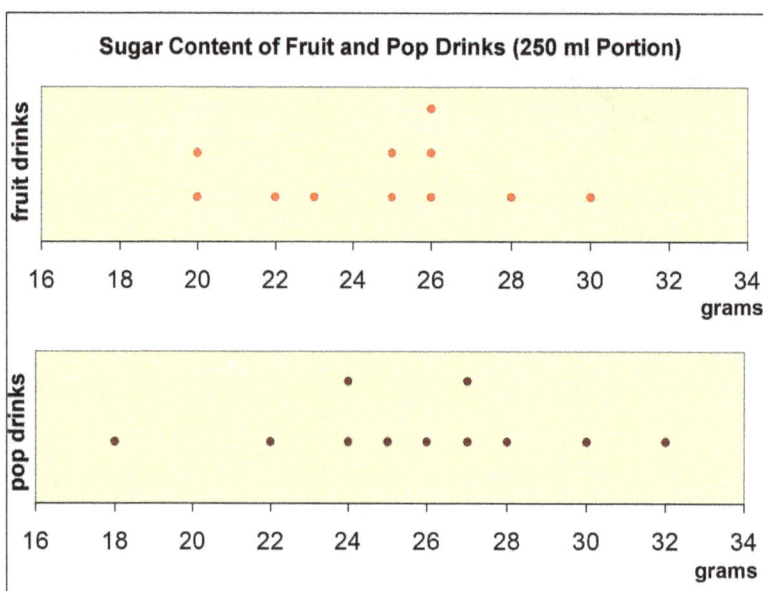

Sugar Content of Fruit and Pop Drinks (250 ml Portion)

   a. Based on the plots, which type of drink, if any, tends to contain more sugar?

   Justify your answer.

   b. Which type of drink, if any, has greater variability in the amount of sugar it has?

   Justify your answer.

10. The map shows the center of Phoenix, Arizona. The distance AC along North 35th Avenue is 2 miles and the distance CB along West McDowell Road is also 2 miles.

Calculate, to the hundredth of a mile, how much longer it is to travel from A to C and then from C to B than to travel from A to B along Grand Avenue.

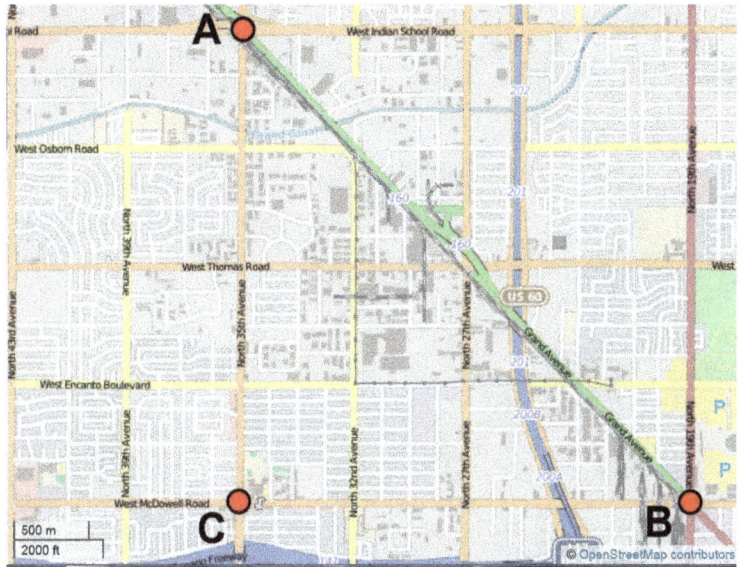

© OpenStreetMap contributors
Licensed under the license at
www.openstreetmap.org/copyright

11. A bag contains 5 blue socks, 2 red socks, and 6 white socks.

a. You draw one randomly. What is the probability that the sock is white?

b. The first sock was indeed white! You draw another one randomly. What is the probability that this one is white?

c. You draw two socks randomly. What is the probability of getting two socks of the same color?

Grade 7 (Pre-algebra) End-of-Year Test

This test is quite long, because it contains lots of questions on all of the major topics covered in the *Math Mammoth Grade 7 Complete Curriculum*. Its main purpose is to be a diagnostic test—to find out what the student knows and does not know about these topics.

You can use this test to evaluate a student's readiness for an Algebra 1 course. In that case, it is sufficient to administer the *first four sections* (Integers through Ratios, Proportions, and Percent), because the topics covered in those are prerequisites for algebra or directly related to algebra. The sections on geometry, statistics, and probability are not essential for a student to be able to continue to Algebra 1. The Pythagorean Theorem is covered in high school algebra and geometry courses, so that is why it is not essential to master, either.

Since the test is so long, I recommend that you break it into several parts and administer them on consecutive days, or perhaps on morning/evening/morning/evening. Use your judgment.

A calculator is *not* allowed for the first three sections of the test: Integers, Rational Numbers, and Algebra.

A <u>basic</u> calculator *is* allowed for the last five sections of the test: Ratios, Proportions, and Percent; Geometry, The Pythagorean Theorem, Probability, and Statistics.

The test is evaluating the student's ability in the following content areas:

- operations with integers
- multiplication and division of decimals and fractions, including with negative decimals and fractions
- converting fractions to decimals and vice versa
- simplifying expressions
- solving linear equations
- writing simple equations and inequalities for word problems
- graphs of linear equations
- slope of a line
- proportional relationships and unit rates
- basic percent problems, including percentage of change
- working with scale drawings
- drawing triangles
- the area and circumference of a circle
- basic angle relationships
- cross-sections formed when a plane cuts a solid
- solving problems involving area, surface area, and volume
- using the Pythagorean Theorem
- simple probability
- listing all possible outcomes for a compound event
- experimental probability, including designing a simulation
- biased vs. unbiased sampling methods
- making predictions based on samples
- comparing two populations and determining whether the difference in their medians is significant

If you are using this test to evaluate a student's readiness for Algebra 1, I recommend that the student score a minimum of 80% on the first four sections (Integers through Ratios, Proportions, and Percent). The subtotal for those is 118 points. A score of 94 points is 80%.

I also recommend that the teacher or parent review with the student any content areas in which the student may be weak. Students scoring between 70% and 80% in the first four sections may also continue to Algebra 1, depending on the types of errors (careless errors or not remembering something, versus a lack of understanding). Use your judgment.

You can use the last four sections to evaluate the student's mastery of topics in Math Mammoth Grade 7 Curriculum. However, mastery of those sections is not essential for a student's success in an Algebra 1 course.

My suggestion for points per item is as follows.

Question #	Max. points	Student score
Integers		
1	2 points	
2	2 points	
3	3 points	
4	6 points	
5	2 points	
6	3 points	
subtotal		/ 18
Rational Numbers		
7	8 points	
8	3 points	
9	3 points	
10	2 points	
11	4 points	
subtotal		/ 20
Algebra		
12	6 points	
13	3 points	
14	12 points	
15	2 points	
16a	1 point	
16b	2 points	
17	3 points	
18	4 points	

Question #	Max. points	Student score
19a	2 points	
19b	1 point	
20	8 points	
21	2 points	
22a	2 points	
22b	1 point	
subtotal		/ 49
Ratios, Proportions, and Percent		
23	4 points	
24a	1 point	
24b	2 points	
24c	1 point	
24d	1 point	
25a	1 point	
25b	2 points	
26	2 points	
27	2 points	
28a	2 points	
28b	2 points	
29	2 points	
30	2 points	
31	2 points	
32	Proportion: 1 point Solution: 2 points	
33	2 points	
subtotal		/ 31
SUBTOTAL FOR THE FIRST FOUR SECTIONS:		**/118**

Question #	Max. points	Student score
Geometry		
34a	2 points	
34b	2 points	
35	3 points	
36	2 points	
37	2 points	
38	2 points	
39a	1 points	
39b	3 points	
40a	2 points	
40b	2 points	
41	2 points	
42	3 points	
43a	2 points	
43b	2 points	
44a	2 points	
44b	2 points	
45a	2 points	
45b	1 point	
46a	1 point	
46b	2 points	
subtotal		/ 40
The Pythagorean Theorem		
47	2 points	
48	2 points	
49	2 points	
50	3 points	
subtotal		/9

Question #	Max. points	Student score
Probability		
51	3 points	
52a	2 points	
52b	1 point	
52c	1 point	
52d	1 point	
53	3 points	
54	3 points	
subtotal		/14
Statistics		
55	2 points	
56a	1 point	
56b	2 points	
56c	2 points	
57	2 points	
58a	1 point	
58b	1 point	
58c	1 point	
58d	3 points	
subtotal		/15
SUBTOTAL FOR THE LAST FOUR SECTIONS:		**/78**
TOTAL		**/196**

Math Mammoth End-of-Year Test - Grade 7

Integers

A calculator is not allowed for the problems in this section.

1. Give a real-life situation for the sum $-15 + 10$.

2. Give a real-life situation for the product $4 \cdot (-2)$.

3. Represent the following operations on the number line.

a. $-1 - 4$

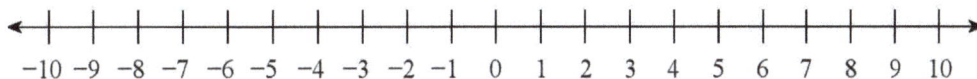

b. $-2 + 7$

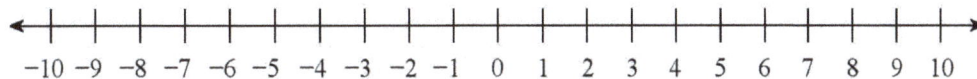

c. $-2 + (-7)$

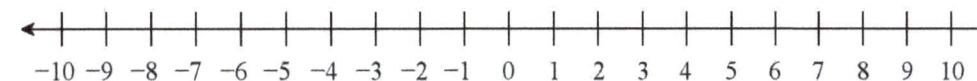

4. Solve.

a. $-13 + (-45) + 60 =$ _____	**b.** $-8 - (-7) =$ _____	**c.** $2 - (-17) + 6 =$ _____
d. $-3 \cdot (-8) =$ _____	**e.** $48 \div (-4) =$ _____	**f.** $(-2) \cdot 3 \cdot (-2) =$ _____

5. The expression $|20 - 31|$ gives us the distance between the numbers 20 and 31.
   Write a similar expression for the distance between -5 and -15 and simplify it.

6. Divide. Give your answer as a fraction or mixed number in lowest terms.

a. $1 \div (-8)$	**b.** $-4 \div 16$	**c.** $-21 \div (-5)$

Rational Numbers

A calculator is not allowed for the problems in this section.

7. Multiply and divide. For problems with fractions, give your answer as a mixed number in lowest terms.

a. $-\dfrac{2}{7} \cdot \left(-3\dfrac{5}{8}\right)$	**b.** $27.5 \div 0.6$
c. $-0.7 \cdot 1.1 \cdot (-0.001)$	**d.** $(-0.12)^2$
e. $\dfrac{\dfrac{3}{4}}{\dfrac{5}{12}}$	**f.** $\dfrac{5\frac{1}{2}}{-\dfrac{7}{8}}$
g. $-\dfrac{1}{6} \cdot 1.2$	**h.** $-\dfrac{2}{5} \div (-0.1)$

8. Write the decimals as fractions.

a. 0.1748	**b.** −0.00483	**c.** 2.043928

9. Write the fractions as decimals.

a. $-\dfrac{28}{10,000}$	**b.** $\dfrac{2,493}{100}$	**c.** $7\dfrac{1338}{100,000}$

10. Convert to decimals. If you find a repeating pattern, give the repeating part. If you don't, round your answer to five decimals.

a. $\dfrac{7}{13}$	**b.** $1\dfrac{9}{11}$

11. Give a real-life context for each multiplication or division. Then solve.

a. $1.2 \cdot 25$

b. $(3/5) \div 4$

Algebra

A calculator is not allowed for the problems in this section.

12. Simplify the expressions.

a. $7s + 2 + 8s - 12$	**b.** $x \cdot 5 \cdot x \cdot x \cdot x$	**c.** $3(a + b - 2)$
d. $0.02x + x$	**e.** $1/3(6w - 12)$	**f.** $-1.3a + 0.5 - 2.6a$

13. Factor the expressions (write them as multiplications).

a. $7x + 14$ $=$	**b.** $15 - 5y$ $=$	**c.** $21a + 24b - 9$ $=$

14. Solve the equations.

a. $\quad 2x - 7 \;=\; -6$	**b.** $\quad 2 - 9 \;=\; -z + 4$
c. $\quad 120 \;=\; \dfrac{c}{-10}$	**d.** $\quad 2(x + \frac{1}{2}) \;=\; -15$
e. $\quad \dfrac{2}{3}x \;=\; 266$	**f.** $\quad x + 1\dfrac{1}{2} \;=\; \dfrac{3}{8}$

144

15. Chris can run at a constant speed of 12 km/h. How long will it take him to run from his home to the park, a distance of 0.8 km?

Remember to check that your answer is reasonable.

16. **a.** Which equation matches the situation?

A pair of binoculars is discounted by 1/5 of their original price (p), and now they cost $48.

| $\frac{p}{5} = 48$ | $\frac{4p}{5} = 48$ | $\frac{5p}{4} = 48$ | $p - 1/5 = 48$ | $p - 4/5 = 48$ | $5p - 4 = 48$ |

b. Solve the equation to find the original price of the binoculars.

17. The perimeter of a rectangle is 254 cm. Its length is 55 cm. Represent the width of the rectangle with a variable and write an equation to solve for the width. Then solve your equation.

18. Solve the inequalities and plot their solution sets on a number line. Write appropriate multiples of ten under the bolded tick marks (for example, 30, 40, and 50).

a. $3x - 7 < 83$

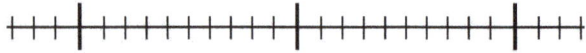

b. $2x - 16.3 \geq 10.5$

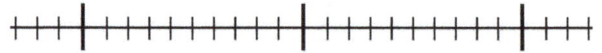

19. You need to buy canning jars. They cost $15 per box, and you only have $150 to spend. You also have a coupon for a $25 discount on your total. How many boxes can you buy at most?

a. Write an inequality for the problem and solve it.

b. Describe the solution of the inequality in words.

20. *Solve.

a. $\quad 9y - 2 + y \quad = \quad 5y + 10$	**b.** $\quad 2(x + 7) \quad = \quad 3(x - 6)$
c. $\quad \dfrac{y + 6}{-2} \quad = \quad -10$	**d.** $\quad \dfrac{w}{2} \; - \; 3 \; = \; 0.8$

21. *Draw a line that has a slope of 1/2 and that goes through the point (0, 4).

22. **a.** *Draw the line $y = -2x + 1$.

    b. *What is its slope?

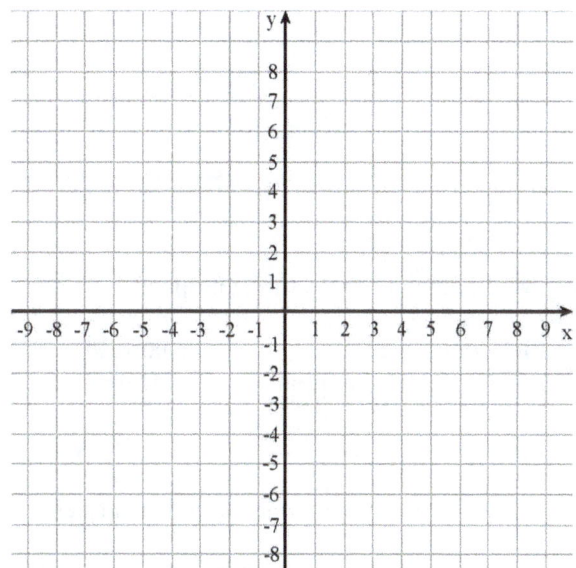

You may use a basic calculator for all the problems in this section.

23. (1) Write a unit rate as a complex fraction. (2) Then simplify it. Be sure to include the units.

---

a. Lily paid $6 for 3/8 lb of nuts.

---

b. Ryan walked 2 ½ miles in 3/4 of an hour.

---

24. The graph below shows the distance covered by a moped advancing at a constant speed.

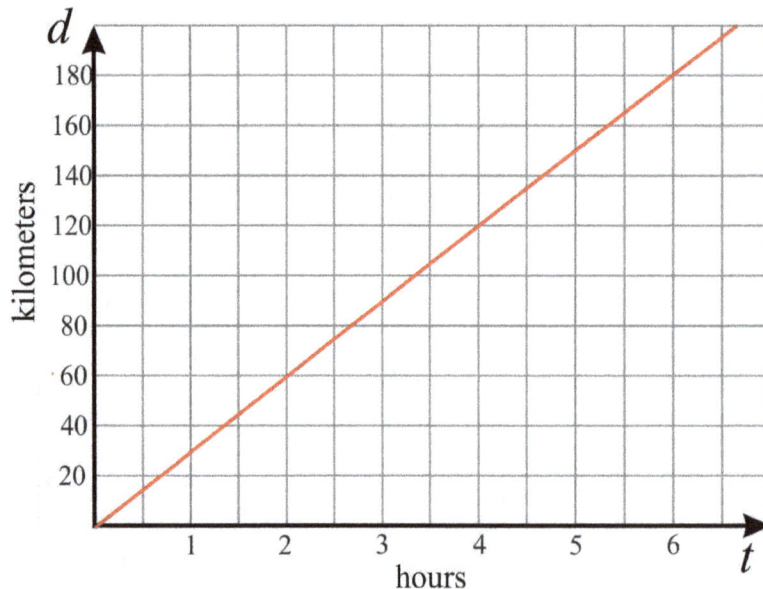

    a. What is the speed of the moped?

    b. Plot on the line the point that corresponds to the time $t = 4$ hours.
       What does that point signify in this context?

    c. Write an equation relating the quantities d and t.

    d. Plot the point that corresponds to the unit rate in this situation.

25. A Toyota Prius is able to go 565 miles on 11.9 gallons of gasoline (highway driving). A Honda Accord can travel 619 miles on 17.2 gallons of gasoline (highway driving). (Source: Fueleconomy.gov)

   a. Which car gets better gas mileage?

   b. Calculate the difference in costs if you drive a distance of 300 miles with each car, if gasoline costs $3.19 per gallon.

26. Sally deposits $2,500 at 8% interest for 3 years. How much can she withdraw at the end of that period?

27. A ticket to a fair initially costs $10. The price is increased by 15%. Then, the price is decreased by 25% (from the already increased price). What is the final price of the ticket?

28. In December, Sarah's website had 72,000 visitors. In December of the previous year it had 51,500 visitors.

   a. Find the percentage of increase to the nearest tenth of a percent in the number of visitors her website had for that year.

   b. If the number of visitors continues to grow at the same rate, about how many visitors (to the nearest thousand) will her site have in December of the following year?

29. Alex measured the rainfall on his property to be 10.5 cm in June, which he calculated to be a 35% increase compared to the previous month. How much had it rained in May?

30. A square with sides of 15 cm is enlarged in a ratio of 3:4. What is the area of the resulting square?

31. How long is a distance of 8 km if measured on a map with a scale of 1:50,000?

32. Write a proportion for the following problem and solve it.

600 ml of oil weighs 554 g.
How much would 5 liters of oil weigh?   ——————— = ———————

33. A farmer sells potatoes in sacks of various weights. The table shows the price per weight.

Weight	5 lb	10 lb	15 lb	20 lb	30 lb	50 lb
Price	$4	$7.50	$9	$12	$15	$25

a. Are these two quantities in proportion?

Explain how you can tell that.

b. If so, write an equation relating the two and state the constant of proportionality.

Geometry

You may use a basic calculator for all the problems in this section.

34. The rectangle you see below is Jayden's room, drawn here at a scale of 1:45.

   a. Calculate the area of Jayden's room in reality, in square meters.
   Hint: measure the dimensions of the rectangle in centimeters.

   b. Reproduce the drawing at a scale of 1:60.

Scale 1:45

35. A room measures 4 ¼ in. by 3 ½ in. in a house plan with a scale of 1 in : 3 ft.
    Calculate the actual dimensions of the room.

36. Calculate the area of a circle with a diameter of 16 cm.

37. Calculate the circumference of a circle with a radius of 9 inches.

38. Draw a triangle with sides 8 cm, 11 cm, and 14.5 cm using a compass and a ruler.

39. A triangle has angles that measure 36°, 90°, and 54°, and a side of 8 cm.

    a. Does the information given determine a unique triangle?

    b. If so, draw the triangle. If not, draw several different triangles that fit the description.

40. **a.** Write an equation for the measure of angle x, and solve it.

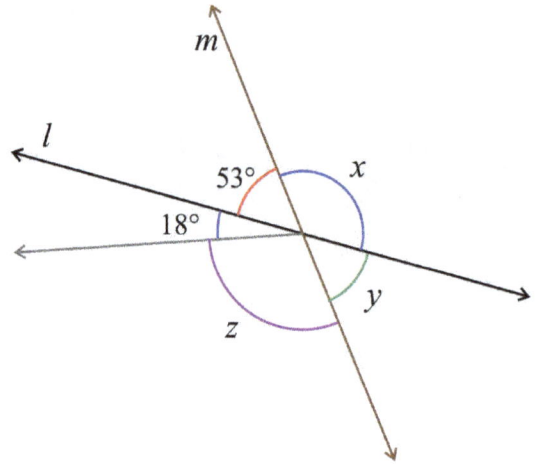

m

l

53°

x

18°

y

z

b. Write an equation for the measure of angle z, and solve it.

41. Calculate the measure of the unknown angle x.

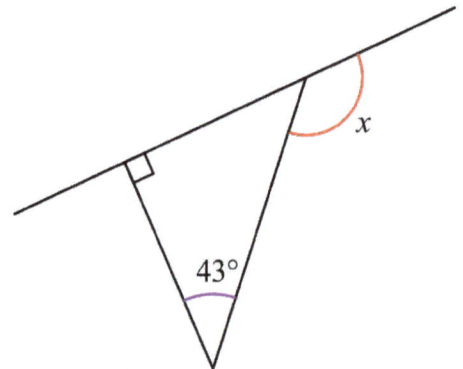

x

43°

42. Describe the cross sections formed by the intersection of the plane and the solid.

a.

The cross section is

_____.

b.

The cross section is

_____.

c.

The cross section is

_____.

43. **a.** Calculate the volume enclosed by
    the roof (the top part).

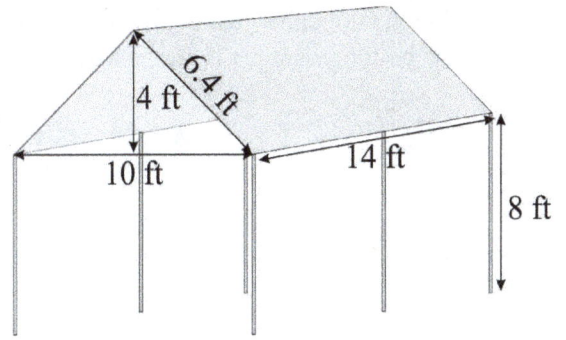

b. Calculate the total volume enclosed by the canopy.

44. Two identical trapezoids are placed inside a 15 cm by 15 cm square.

a. Calculate their area.

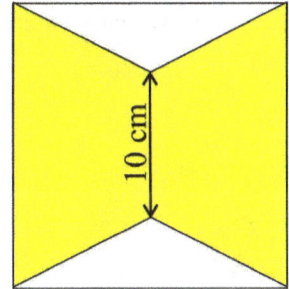

b. What percentage of the square do the trapezoids cover?

45. **a.** *Find the volume of the cylindrical part of the juicer, if its bottom diameter is 12 cm and its height is 4.5 cm.

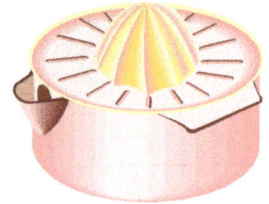

b. *Convert the volume to milliliters and to liters, considering that 1 ml = 1 cm^3.

46. **a.** *How many cubic inches are in one cubic foot?

b. *The edges of a cube measure 3 ¼ ft. Calculate the volume of the cube in cubic inches.

The Pythagorean Theorem

You may use a basic calculator for all the problems in this section.

47. ***a.** What is the area of a square, if its side measures $\sqrt{5}$ m?

    *b. How long is the side of a square with an area of 45 cm^2?

48. ***Determine whether the lengths 57 cm, 95 cm, and 76 cm form a right triangle. Show your work.**

49. ***Solve for the unknown side of the triangle to the nearest tenth of a centimeter.**

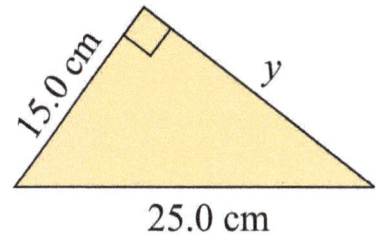

50. *You and your friends are at a river at point A. You suddenly remember you need something from home, which is at point C. So you decide to go home (distance AC) and then walk along the road (distance CB) to meet your friends, who will walk along the riverside from A to B.

If ABC is a right triangle, AC = 120 m, and CB = 110 m, how much longer distance (in meters) will you walk than your friends?

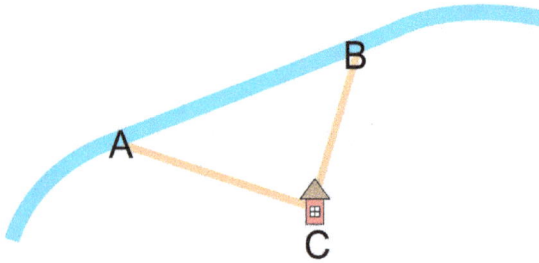

Probability

You may use a basic calculator for all the problems in this section.

51. You randomly pick one marble from these marbles.
    Find the probabilities:

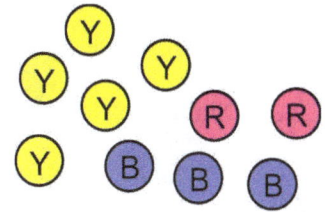

    a. P(not red)

    b. P(blue or red)

    c. P(green)

52. A cafeteria offers a main dish with chicken or beef. The customer then chooses a portion of rice, pasta, or potatoes, and a side dish of green salad, green beans, steamed cabbage, or coleslaw.

    a. Draw a tree diagram or make a list of all the possible meal combinations.

    A customer chooses the parts of the meal randomly. Find the probabilities:

    b. P(beef, rice, coleslaw)

    c. P(no coleslaw nor steamed cabbage)

    d. P(chicken, green salad)

53. John and Jim rolled a die 1,000 times. The bar graph shows their results. Based on the results, which of the following conclusions, if any, are valid?

   (a) This die is unfair.

   (b) On this die, you will always get more 1s than 6s.

   (c) Next time you roll, you will not get a 4.

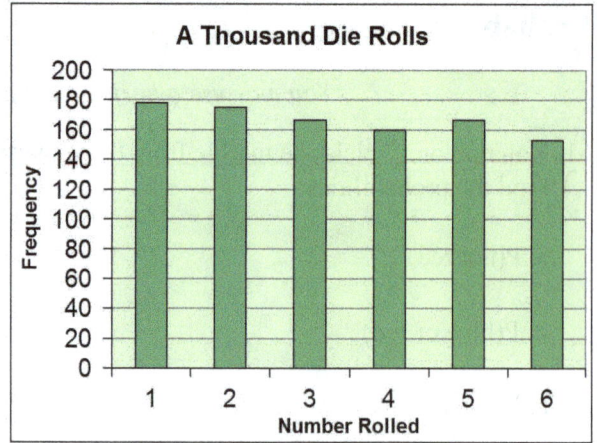

A Thousand Die Rolls

(Frequency vs. Number Rolled)

54. Let's assume that when a child is born, the probability that it is a boy is 1/2 and also 1/2 for a girl. One year, there were 10 births in a small community, and nine of them were girls. Explain how you could use coin tosses to simulate the situation, and to find the (approximate) probability that out of 10 births, exactly nine are girls. (You do not have to actually perform the simulation—just explain how it would be done.)

You may use a basic calculator for all the problems in this section.

55. To determine how many students in her college use a particular Internet search engine, Cindy chose some students randomly from her class, and asked them whether they used that search engine.

    Is Cindy's sampling method biased or unbiased?

    Explain why.

56. Four people are running for mayor in a town of about 20,000 people. Three polls were conducted, each time asking 150 people who they would vote for. The table shows the results.

	Clark	Taylor	Thomas	Wright	Totals
Poll 1	58	19	61	12	150
Poll 2	68	17	56	9	150
Poll 3	65	22	53	10	150

    a. Based on the polls, predict the winner of the election.

    b. Assuming there will be 8,500 voters in the actual election, estimate to the nearest hundred votes how many votes Thomas will get.

    c. Gauge how much off your estimate might be.

57. Gabriel randomly surveyed some households in a small community to determine how many of them support building a new highway near the community. Here are the results:

If the community contains a total of 2,120 households, predict how many of them would support building the highway.

Opinion	Number
Support the highway	45
Do not support it	57
No opinion	18

58. Researchers compared two different methods for losing weight by assigning 50 overweight people to use each method. The side-by-side boxplots show how many pounds people in each group lost.

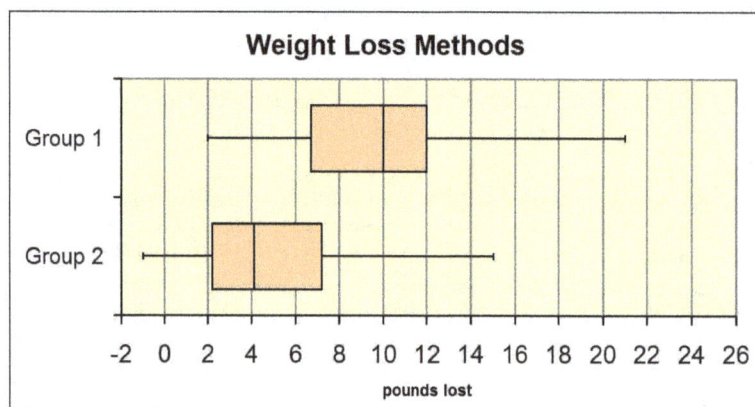

a. Just looking at the two distributions, which group, if any, appears to have lost more weight?

b. Which group, if any, appears to have a greater variability in the amount of weight lost?

c. In Group 2, there is one person whose weight loss was −1 pound. What does that mean?

d. Is one of the weight loss methods significantly better than the other?

If so, which one?

Justify your reasoning.

Math Mammoth Grade 7 Review Workbook Answers

The Language of Algebra Review, p. 7

1. a. 2,700   b. 9   c. 7 1/2

2. The associative property of multiplication.

3. a. 50   b. 8/67

4. a. $l_1 = l_2 - 1.5$

   b. $\dfrac{5w}{6} = 23$

5. No. For example, $5 - 3$ is not the same as $3 - 5$.

6. a. $p - p/5$ or $p - (1/5)p$ or $(1 - 1/5)p$ or $(4/5)p$ or $0.8p$
   b. $0.18y$
   c. $3n + 8m$

7. a. $3x + 2$
   b. $2x^4$
   c. $20v$
   d. $96v^2$
   e. $36z^3$
   f. $2f + 6g$

8. a.

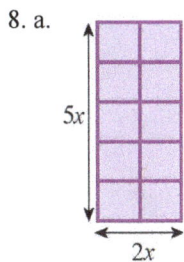

   b. The area is $10x^2$.
   c. The perimeter is $14x$.

9. a. $12v - 108$

   b. $3a + 3b + 6$

   c. $1.5t - 3x$

10.

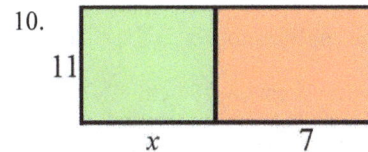

11.

Expression	The terms in it	Coefficient(s)	Constants
a^8	a^8	1	none
$2x + 9y$	$2x$ and $9y$	2 and 9	none

12. The side is $6s + 9$.

13. a. $48x + 12 = 12(4x + 1)$

   b. $40x - 25 = 5(8x - 5)$

   c. $6y - 2z = 2(3y - z)$

   d. $56t - 16s + 8 = 8(7t - 2s + 1)$

The Language of Algebra Test, p. 9

1. $2s^2 + 5t + 9$

2. a. $2(7 - 2)^2 = 2(5)^2 = \underline{50}$
   b. $\dfrac{1}{6} + \dfrac{6+1}{3} = \dfrac{15}{6} = 2\dfrac{1}{2}$

3. The distributive property of multiplication over addition states that $a(b + c) = ab + ac$.

4. These two rectangles show that $5(z + 4) = 5z + 20$

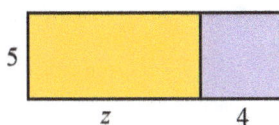

5. a. $7x + 21 = 7(x + 3)$
   b. $24k + 80 = 8(3k + 10)$

6. a. $4v + 5$

 b. $5v^4$

 c. $5x + 3$

7. a. $7(x - 1) = 14$; $x = 3$

 b. $x^2 - 2 = 23$; $x = 5$

8. a. $\$50 - x \cdot \$3$, which is usually written as $50 - 3x$.

 b. $\dfrac{9}{10}p$ or $0.9p$

9. a. $4x \cdot 5x + 2x \cdot 3x$, which simplifies to $20x^2 + 6x^2 = 26x^2$

 b. The area is $26(2 \text{ cm})^2 = 26 \cdot 4 \text{ cm}^2 = 104 \text{ cm}^2$

 c. $4x + 5x + 2x + 3x + 2x + x + 5x = 22x$

Integers Review, p. 11

1. a. A ball was dropped from 18 ft above sea level; it fell 12 ft. Now the ball is at _6_ ft.

 b. John had a \$12 debt. He earned \$18. Now he has _\$6_.

 c. John had \$12. He had to pay his dad \$18. Now he has _−\$6_.

 d. A diver was at the depth of 18 ft. Then he rose 12 ft. Now he is at _−6_ ft.

 e. The temperature was −12°C and fell 18°C. Now it is _−30_ °C.

 c. $12 - 18 = \underline{-6}$

 b. $-12 + 18 = \underline{\ 6}$

 d. $-18 + 12 = \underline{-6}$

 e. $-12 - 18 = \underline{-30}$

 a. $18 - 12 = \underline{\ 6}$

2. a. $5°C > -12°C$ or $-12°C < 5°C$

 b. $-\$250 > -\$400$ or $-\$400 < -\$250$

3. a. -13 b. -11 c. -5

4. Answers will vary. Please check the student's work. For example:

a. $2 + (-4) = -2$ $-11 + 9 = -2$	b. $13 + (-13) = 0$ $-24 + 24 = 0$	c. $8 + (-5) = 3$ $-3 + 6 = 3$

5. a. $|-14 \text{ m}| = 14 \text{ m}$
   Here, the absolute value shows the distance from the shark to the surface of the water.

 b. $|-\$31| = \$31$
   Here, the absolute value shows the size of Shelley's debt.

6. a. -8 b. $-(-100) = 100$ c. $-(2 + 5) = -7$ d. $|-45| = 45$

7. a. Now it is 2°C.

 b. Now it is −6°C.

 c. Now it is −6°C.

 d. Now it is −11°C.

 $^-2 + 4 = 2$

 $-11 + 5 = -6$

 $2 + (-8) = -6$

 $-3 - 8 = -11$

8.

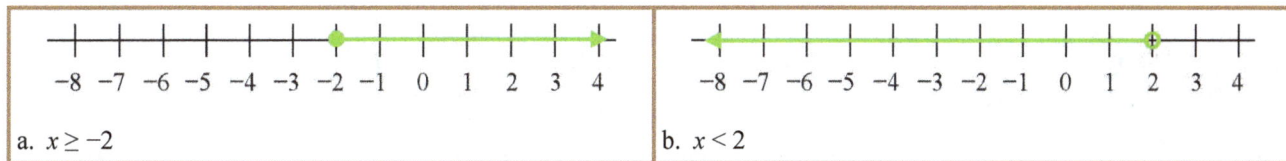

| a. $x \geq -2$ | b. $x < 2$ |

9. a. -3 b. -17

10. The total electric charge of this ion is $-e$ (or $-1e$).

11. The expression for the distance can be written as $|-6 \text{ m} - (-8 \text{ m})|$ or $|-8 \text{ m} - (-6 \text{ m})|$. It can be written even without absolute value, as $-6 \text{ m} - (-8 \text{ m})$ because we know which number is greater and we can subtract the smaller number (-8 m) from the greater (-6 m). Each one of those will give the correct distance, 2 m. If we were dealing with variable(s), the distance would need to be expressed as the absolute value of the difference of the two quantities.

12. a. -27 b. 3   c. 290

13. a. $1 + 7 = 8$
    b. $2 + 11 = 13$
    c. $-20 + 6 = -14$
    d. $3 + (-8) = -5$

14. Xerxes ruled 1 year longer than Darius II.
    Xerxes I of Persia: $486 - 465 \text{ BC} = 21$ years.
    Darius II of Persia: $424 - 404 \text{ BC} = 20$ years.

15.

| a. $-2 \cdot (-4) = 8$
 $-2 \cdot 4 = -8$ | b. $(-3) \cdot (-8) = 24$
 $7 \cdot (-12) = -84$ | c. $(-3) \cdot 3 \cdot (-1) = 9$
 $-7 \cdot (-2) \cdot (-2) = -28$ |

16. a. True
    b. False. For example, -14 is less than -6, yet its absolute value, 14, is greater than 6.
    c. True.
    d. False. For example, for -6, the absolute value of its opposite is 6, but the opposite of its absolute value is -6.

17.

| a. $-10 \div (-5) = 2$
 $24 \div (-3) = -8$ | b. $(-12) \div (-4) = 3$
 $21 \div (-3) = -7$ | c. $-56 \div 7 = -8$
 $-120 \div (-10) = 12$ |

18.

| a. $-5 \cdot \mathbf{6} = -30$ | b. $2 \cdot (\mathbf{-9}) = -18$ | c. $-8 \cdot (\mathbf{-6}) = 48$ |
| d. $-42 \div (\mathbf{-7}) = 6$ | e. $-64 \div \mathbf{8} = -8$ | f. $81 \div (\mathbf{-9}) = -9$ |

19. a. $y = -20$ b. $b = 12$ c. $y = -7$

20. Answers will vary. Please check the student's work. For example:
    $3 \cdot (-10) = (-10) + (-10) + (-10) = -30$.
    Each of three trees had birds perched in them. Ten birds flew away from each tree. The total loss was 30 birds.

21. a. $-\dfrac{1}{6}$ b. $-\dfrac{1}{5}$ c. $\dfrac{6}{7}$

22. a. 9   b. 60   c. 1

1. $-5 + 7 = 2$

2. Answers will vary. Please check the student's answer. For example: $-8 + 8 = 0$

3. a. 3   b. 19   c. -19 d. -3

4. a. -21 b. -26 c. -146 d. -27

5. Answers will vary. Please check the student's explanation. For example, a submarine was cruising at 80 m below the surface. Then it sank 30 m deeper. Where is it now? -80 m $- 30$ m $= -110$ m.

6. $-21 + 15 - 35 + 50 = \$9$. Now his balance is \$9.

7. a. 14   b. -4 c. 9   d. -5

8. She did eleven points better.

9. a. The distance is $|-2 - (-18)|$, which simplifies to $|16| = 16$.

   b. $|x - 5|$

   c. $|-2 - 5| = |-7| = 7$

10.

a. $-7 \cdot \underline{5} = -35$	b. $2 \cdot \underline{(-50)} = -100$	c. $-50 \div \underline{5} = -10$

11.

a. $15 \div (-3) = \underline{-5}$	b. $-2 \div (-10) = 2/10 = \underline{1/5}$
c. $-20 \div 6 = -10/3 = \underline{-3\ 1/3}$	d. $72 \div (-48) = -6/4 = \underline{-1\ 1/2}$

12. a. $(-3)(4) - 2 = -12 - 2 = \underline{-14}$

   b. $(-3)^2 + 1 = 9 + 1 = \underline{10}$

   c. $-2(4 - 5) = (-2)(-1) = \underline{2}$

13. a. $y = -12$
   b. $a = 15$
   c. $w = -6$

1. a. 27 cm^3 b. 2.25 in^2 c. 1000 m^3

2. $(2s)^3 = 8s^3$

3. Volume $= 8s^3 = 8 \cdot (1.7 \text{ cm})^3 = 39.304 \text{ cm}^3$

4. a. 201 decreased by a number is 167.

   b. The product of a number and 7 equals 7/24.

Equation	Solution
$201 - x = 167$	$x = 201 - 67 = 34$
$7b = 7/24$	$b = 1/24$

5. No, it isn't. For example, $2 - 5$ is not equal to $5 - 2$.

6. The associate property of addition.

7.

Expression	the terms in it	coefficient(s)	Constantss
$(5/6)s^2$	$(5/6)s^2$	5/6	none
$x + 2y + 8$	x, $2y$, and 8	2	8
$p \cdot 46$	$p \cdot 46$	46	none

8.

$3(2x + 7)$ and

$3 \cdot 2x + 3 \cdot 7 = 6x + 21$

9.

a. $3x + 6 = 3(x + 2)$	b. $10z - 20 = 10(z - 2)$

10. a. $-|2| = -2$ b. $|-2| = 2$

11. a. $d \leq -2$ ft

    b. We consider her balance as negative: $m \geq -\$100$.

    c. $t \geq -20°C$

12. $(3/4)x$

13. $3x + 21 = 3(x + 7)$

14. a. $2t + 9$

    b. $4x^3$

    c. $10xy^2$

1.

a. $\dfrac{7}{8} \cdot 4 = \dfrac{28}{8} = 3\dfrac{1}{2}$	b. $5 \cdot \dfrac{2}{10} + 1 = \dfrac{10}{10} + 1 = 2$	c. $\dfrac{10 + 3}{8 - 1} = \dfrac{13}{7} = 1\dfrac{6}{7}$

2.

a. $\dfrac{23}{19} = 1\dfrac{4}{19}$	b. $3 \cdot 10^2 - 2 \cdot 3^2 = 300 - 18 = \underline{282}$

3. It is the associative property of multiplication.

4. a. Choose two variables to denote the lengths of the two broomsticks.
   Let x be the length of the wooden broomstick.
   Let y be the length of the metal one.

   b. $x = y + 20$ or $x - 20 = y$.

5. a. $3(p - \$5) = \$16.80$
   b. One sun hat would have cost \$10.60 originally.
   If the customer had bought them at the normal price, the cost would have been $\$16.80 + 3 \cdot \$5 = \$31.80$.
   Now divide that by 3 to get the price of one hat without a discount: $\$31.80 \div 3 = \$10.60$.

6.

a. $11p + 1$	b. $42p^2$	c. $12f^4$

7.

a. $\lvert -71 \rvert = 71$	b. $-\lvert -2 \rvert = -2$	c. $\lvert -9 + 5 \rvert = 4$	d. $-(-84) = 84$
e. $\lvert -9 \rvert + \lvert -5 \rvert = 14$		f. $\lvert -9 \rvert - \lvert 5 \rvert = 4$	

8. a. $h \geq 200$ ft
   b. $d < -\$120$, where d is Liz's account balance. Or, $s > \$120$ where s is the amount of debt.
   c. $x \leq 8$
   d. $x \leq 120$ cm

9.

a. $5 - 7 = {}^-2$	c. $2 + ({}^-6) = {}^-4$	e. ${}^-30 + ({}^-10) = {}^-40$	g. ${}^-51 + 51 = 0$
b. ${}^-1 - 18 = {}^-19$	d. $4 - ({}^-1) = 5$	f. $0 - 49 = {}^-49$	h. ${}^-9 + 2 = {}^-7$

10.

   Step 1   2   3   4   5

   b. The pattern grows by adding one flower across the top, and one flower on each side.

   c. There will be $3 \cdot (39 + 1) = 120$ flowers in step 39.

   d. $3n + 3$ or $3(n + 1)$

Solving One-Step Equations Review, p. 20

1. a. $x = -13$ b. $x = 4$ c. $x = 10$ d. $z = 9$
   e. $x = -132$ f. $q = 120$ g. $c = -1,000$ h. $a = -105$

2. Equation: $3p = 837$. Solution: $p = 279$. One solar panel cost $279.

3. Equation: $s/7 = 187$. Solution: $s = 1,309$. Andrew's salary was $1,309.

4. Substituting the values given in the problem in the formula $d = vt$ gives us the equation 1.2 km $= 20$ km/h $\cdot t$.
   Solution: $t = 1.2$ km $/ (20$ km/h$) = 0.06$ h $= 3.6$ min $= 3$ min 36 sec.

5. $v = d/t = 1.5$ km/3 min $= 1.5$ km/(3/60 h) $= 1.5$ km/(1/20 h) $= 30$ km/h.

6. The first half: $t = 1$ mi/12 mph $= (1/12)$ h $= 5$ minutes. The second half: $t = 1$ mi/15 mph $= (1/15)$ h $= 4$ minutes.
   It will take Ed $5 + 4 = 9$ minutes to get to school.

Solving One-Step Equations Test, p. 22

1.

<table>
<tr>
<td>

a.
$$x + 8 = -13 \qquad |-8$$
$$x = \mathbf{-21}$$
Check:
$$-21 + 8 \overset{?}{=} -13$$
$$-13 = -13 \quad \checkmark$$

</td>
<td>

b.
$$4 - (-2) = -y$$
$$6 = -y \qquad |\cdot(-1)$$
$$y = \mathbf{-6}$$
Check:
$$4 - (-2) \overset{?}{=} -(-6)$$
$$6 = 6 \quad \checkmark$$

</td>
</tr>
<tr>
<td>

c.
$$18 - x = -1 \qquad |-18$$
$$-x = -19$$
$$x = \mathbf{19}$$
Check:
$$18 - 19 \overset{?}{=} -1$$
$$-1 = -1 \quad \checkmark$$

</td>
<td>

d.
$$2 - 6 = -z + 5$$
$$-4 = -z + 5 \qquad |-5$$
$$-9 = -z$$
$$z = \mathbf{9}$$
Check:
$$2 - 6 \overset{?}{=} -9 + 5$$
$$-4 = -4 \quad \checkmark$$

</td>
</tr>
<tr>
<td>

e.
$$\frac{x}{10} = -17 + 5$$
$$\frac{x}{10} = -12 \qquad |\cdot 10$$
$$x = \mathbf{-120}$$
Check:
$$\frac{-120}{10} \overset{?}{=} -17 + 5$$
$$-12 = -12 \quad \checkmark$$

</td>
<td>

f.
$$-13 = \frac{c}{-7} \qquad |\cdot(-7)$$
$$91 = c$$
$$c = \mathbf{91}$$
Check:
$$-13 \overset{?}{=} \frac{91}{-7}$$
$$-13 = -13 \quad \checkmark$$

</td>
</tr>
</table>

169

Solving One-Step Equations Test, cont.

2. a. Let p be the price of one pound of chicken. Equation: $7p = \$32.41$. Solution: $p = \$4.63$.

   b. Let x be the weight of Bill's suitcase. Equation: $x + 4.6 = 28.7$ or $28.7 - 4.6 = x$. Solution: $x = 24.1$ kg.

3.
$$d = v \quad t$$
$$\downarrow \quad \downarrow \quad \downarrow$$
600 m = 18 km/h · t

To solve the equation, we need to change 600 m into 0.6 km. To solve the equation, we need to change 600 m into 0.6 km to get 0.6 km = 18 km/h · t. Keeping the units in mind, we can write that as $0.6 = 18t$.

$$0.6 = 18t \qquad | \div 18$$
$$0.0333... = t$$
$$t = 0.0333... \text{ h}$$

To change 0.0333... hours into minutes, multiply it by 60. You will get 2 minutes.

Another way is to use fractions, writing 0.6 as 6/10:

$$6/10 = 18t \qquad | \div 18$$
$$6/180 = t$$
$$t = 1/30 \text{ h}$$

Then of course 1/30 of an hour is 2 minutes.

4. You can travel 21 1/4 kilometers. From the formula $d = vt$, we get $d = 15$ km/h · (1 h 25 minutes)

   = 15 km/h · (1 25/60 h) = 15 · (1 5/12) km = 15 · (17/12) km = 5 · (17/4) km = 85/4 km = 21 1/4 km = 21.25 km

5. a. The airplane averages about 644 mph. The average speed is $d/t = 2900$ mi / (4.5 h) ≈ 644 mph.

   b. The airplane flies at an average of about 1,037 km per hour.
   The average speed is $d/t = 2900$ mi · 1.6 km/mi / (4.5 h) ≈ 1,037 km/h.

Mixed Review 3, p. 24

1. a. $x^2 - 10$ b. $154/k^3$ c. $(x+2)^5$ d. $x + 2^5$

2. a. [square with sides $x+2$]   b. $(x+2)^2$ c. $4(x+2)$ d. $(1.5+2)^2 = 3.5^2 = 12.25$ square units

3.

a. $-2 + 6 = 4$	b. $-3 - 5 = -8$

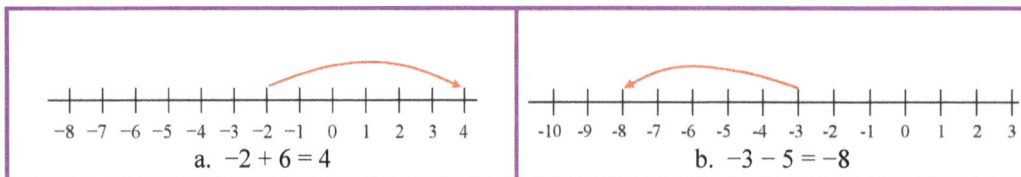

4. The three positive counters cancel out three of the negative counters, which leaves two negative counters. $3 + (-5) = -2$

5. a. $89 + (-35) = 54$ b. $-45 + (-29) = -74$ c. $-78 + 60 = -18$

6.

a. $-2 + (-18)$	b. $56 - (-34)$	c. $-14 + (-24)$	d. $2 + 9$
↓	↓	↓	↓
$-2 - 18 = -20$	$56 + 34 = 90$	$-14 - 24 = -38$	$2 - (-9) = 11$

7. a. $-34°C < -8°C$ b. $-12e < +3e$

8. The commutative property of multiplication is illustrated by the equation $2x = x \cdot 2$.

9.

a. a is 8 and b is 54	b. a is -12 and b is -5
$\|8 - 54\| = \|-46\| = 46$.	$\|-12 - (-5)\| = \|-7\| = 7$.
The distance between 8 and 54 is 46, so the value of the absolute value expression matches that.	The distance between -12 and -5 is 7, so the value of the absolute value expression matches that.

10. Check the student's answer.
    Example: Mark's balance is $-\$15$ and Ann's is $\$15$. This means Mark owes $\$15$ whereas Ann has $\$15$.

11. a. $42s + 28 = 14(3s + 2)$. Another possible answer is $7(6s + 4)$; however the latter is not considered fully factored. Since the numbers 6 and 4 have a common factor of 2, $6s + 4$ can be factored farther to $2(3s + 2)$.

    b. $54z - 18 = 18(3z - 1)$. Another possible answer is $6(9s - 3)$ or $9(6s - 2)$; however the latter expressions are not factored to the lowest terms.

12. a. $x = -175$ b. $y = -7$ c. $z = -7$

13.

a. $2 + 14 = x - 1$ $\quad x = 17$
b. $x^3 = 27$ $\quad x = 3$

14.

a. $(-9) + (-18) = -27$	b. $-21 - (-3) = -18$	c. $17 - 51 = -34$

15. Check the student's answer. For example, Helen has $\$3$. She buys a bag that costs $\$10$. Since she only has $\$3$, she borrows $\$7$ from her mom. Now, Helen owes her mom $\$7$.

16. a. 2   b. 2   c. -2 d. 0

17. a. $5x^2 = 5(-2)^2 = 5(4) = 20$

    b. $-5y + 6 = -5(8) + 6 = -40 + 6 = -34$

    c. $-(y + x) = -(8 + (-2)) = -6$

18. Larry's age is $y - 2$.

19.

Step   1   2   3   4   5

a. Each side grows by one flower.
b. There will be 121 flowers in step 39.
c. Answers will vary. For example, step n will have $3(n + 2) - 2$ flowers. Or, step n will have $3n + 4$ flowers.
   Or, step n will have $n + (n + 2) + (n + 2)$ flowers. Or, step n will have $n + 2(n + 2)$ flowers.

Mixed Review 4, p. 27

1. -4 and 5 because $(-4)^2 - (-4) - 20 = 16 + 4 - 20 = 0$ and $(5)^2 - 5 - 20 = 25 - 5 - 20 = 0$.

2. No, it is not. For example, $24 \div 6$ is not equal to $6 \div 24$.

3.

a. $\quad \dfrac{x}{7} = -15 \qquad \mid \cdot 7$ $\qquad x = -105$	b. $\quad 11 = \dfrac{x}{-12} \qquad \mid \cdot (-12)$ $\quad -132 = x$ $\qquad x = -132$
c. $\quad 7 - x = -3 \qquad \mid -7$ $\qquad -x = -10$ $\qquad x = 10$	d. $\quad 5 \cdot (-8) = -10x$ $\qquad -40 = -10x \qquad \mid \div (-10)$ $\qquad 4 = x$ $\qquad x = 4$

4. a. $\mid -4°C \mid = 4°C$. The absolute value shows how much the temperature is below zero.

   b. $\mid 2{,}500 \text{ ft} \mid = 2{,}500$ ft. The absolute value shows the distance to the surface of the sea.

5. a. Amanda can swim at the speed of $1\ 5/7$ km/h ≈ 1.714 km/h.
   First of all, 35 minutes is $35/60$ hours. Her speed = (distance)/(time) = 1 km / (35/60 h) = $60/35$ km/h
   $= 1\ 25/35$ km/h $= 1\ 5/7$ km ≈ 1.714 km/h

   b. You can walk about 3.4 mph (the exact number is $3\ 9/22 \approx 3.409$).

   In one hour you can walk $15 \cdot 1{,}200$ ft $= 18{,}000$ ft. This is $18{,}000 / 5{,}280 = 3.4090909...$ miles.
   So your speed is about 3.4 miles per hour.

   Here's another way to figure it out: Convert the distance 1,200 ft into miles and the time 4 minutes into hours.

   1200 ft is $1200/5280 = 120/528 = 60/264 = 15/66 = 5/22$ miles. And 4 minutes is $4/60 = 1/15$ hours.

   So your speed is 1200 ft / 4 minutes = (5/22 miles)/(1/15 hours). This is a fraction divided by a fraction, so it becomes

   $$\frac{5}{22} \div \frac{1}{15} = \frac{5}{22} \cdot \frac{15}{1} = \frac{75}{22} = 3\ \frac{9}{22} \text{ miles per hour.}$$

6.

a. $(-3) + (-6) + 5 + 1 = -3$	b. $14 + (-20) + (-31) + 11 = -26$

7. a. $-2\ 2/5$ b. $-2\ 1/5$ c. $2/9$

8. a. $500x = 150{,}000$; $x = 300$. The other side is 300 feet.

   b. One yard is 3 feet, so a square yard is 3 feet by 3 feet = 9 square feet.

   c. $16{,}667$ yd^2

Rational Numbers Review, p. 29

1. Answers will vary. Check the student's answers. For example:
   a. $-3/1$ b. $30/1$ c. $21/100$ d. $-19/10$

2.

172

3. a. $\dfrac{472}{10,000}$	b. $-1\dfrac{2,938,442}{100,000,000}$	c. $2\dfrac{38,166}{100,000}$

4. a. -0.0024 b. 987.2 c. 0.04593

5. a. $0.\overline{21}$ b. $1.099\overline{5}$

6. a. $2.06999...$ b. $0.006812812812...$

7. $0.\overline{7}$ is more. It is $0.0\overline{7}$ more than 0.7.

8. All terminating decimals are rational numbers.

9. All repeating decimals are rational numbers.

10. a. $0.1\overline{36}$ b. 1.60870

11.

a. $a = 6$ and $b = -7$
   Distance: 13
 	
   Absolute value of the difference:
   $\lvert 6 - (-7)\rvert = \lvert 13\rvert = 13$
	b. $a = -1.3$ and $b = -7.6$
	   Distance: 6.3
	
	   Absolute value of the difference:
	   $\lvert -1.3 - (-7.6)\rvert = \lvert 6.3\rvert = 6.3$

12.

a. $0.2 \cdot 0.07 = 0.014$	b. $-0.8 \cdot 0.005 = -0.004$	c. $(-0.2)^3 = -0.008$
d. $-5 \cdot (-2.2) = 11$	e. $-0.2 \cdot 0.1 \cdot (-0.3) = 0.006$	

13. a. $\dfrac{7}{33}$ b. $-2\dfrac{11}{12}$ c. $\dfrac{3}{50}$

14.

a. $1\dfrac{1}{5} \div \left(-\dfrac{1}{4}\right) = \dfrac{6}{5} \cdot \left(-\dfrac{4}{1}\right) = -\dfrac{24}{5} = -4\dfrac{4}{5}$	b. $21 \div 0.06 = 350$

15. a. 60% of $18 = **$10.80**.
   You can multiply $6 \cdot 18 = 108$ in your head and conclude that $0.6 \cdot 18 = 10.8$. Or you can find 10% of $18, which is $1.80, and multiply that by 6.

   b. $\dfrac{1}{4} \cdot 9.6 = **2.4**$. To easily divide 9.6 by 4 in your head, first take half of 9.6, which is 4.8, and then take half of that.

   c. $-0.3 \cdot \dfrac{8}{11} = -\dfrac{3}{10} \cdot \dfrac{8}{11} = -\dfrac{24}{110} = -\dfrac{\mathbf{12}}{\mathbf{55}}$

16. a. $7\dfrac{1}{2}$ b. $2\dfrac{2}{35}$ c. $1\dfrac{1}{3}$

17. a. $1.56 \cdot 0.8 = 1.248$
   Answers will vary. Please check the student's work. For example: What is the discounted price after a $1.56 eraser is discounted by 20%?

   Another example: The sides of a square measure 1.56 m. The sides of another, smaller square are 8/10 of those of the bigger square. How long are the sides of the smaller square?

   b. $6 \div (1/2) = 12$

   Answers will vary. Please check the student's work. For example: How many servings of 1/2 apple do you get from 6 apples?

18. In June, July, and August there are a total of $30 + 31 + 31 = 92$ days. At 16 hours of usage per day, that makes 92 d $\cdot$ 16 h/d = 1472 hours of usage during the 3-month period. The air conditioner uses 2 kW of power, so during those hours it will consume 1472 h $\cdot$ 2 kW = 2944 kWh of energy. Therefore, at \$0.1686 per kWh, the cost to run the air conditioner during those three months will be 2944 kWh $\cdot$ 0.1686 \$/kWh = \$496.36.

19. a. $6.798 \cdot 10^6$ b. $5.6 \cdot 10^{10}$

20. a. 780,000   b. 3,495,800,000

21.

Equation:	Another way: Logical reasoning
$$(7/10)x = 4{,}200 \quad \vert \cdot 10$$ $$7x = 42{,}000 \quad \vert \div 7$$ $$x = 6{,}000$$ The total area of the fire is 6,000 acres.	Since 7/10 of the area is 4,200 acres, we can find 1/10 of the total area by dividing 4,200 by 7. That is 600 acres. Then the total area is 10 times that, or 6,000 acres.

22.

a. $x - \dfrac{2}{9} = 5\dfrac{1}{20} \qquad \vert + 2/9$ $x = 5\dfrac{1}{20} + \dfrac{2}{9}$ $x = 5\dfrac{9}{180} + \dfrac{40}{180}$ $x = 5\dfrac{49}{180}$	a. $5y = -\dfrac{4}{12} \qquad \vert \div 5$ $y = -\dfrac{4}{12} \div 5$ $y = -\dfrac{4}{12} \cdot \dfrac{1}{5}$ $y = -\dfrac{4}{60} = -\dfrac{1}{15}$
c. $\quad 0.94 = 1.1 - x$ $1.1 - x = 0.94 \qquad \vert -1.1$ $-x = -0.16$ $x = 0.16$	d. $\quad -0.3x = 10 \qquad \vert \div (-0.3)$ $x = 10 \div (-0.3)$ $x = -33.\overline{3}$

Rational Numbers Test, p. 34

1.

2. a. $5/4 = 1.25$ b. $-7/10 = -0.7$ c. $9/(-100) = -9/100 = -0.09$

3. a. $5\ 1/1{,}000$ b. $-2\ 482/10{,}000$

4. a. -0.0047 b. 78.7 c. -56.24

5. $0.\overline{6}$ is more; it is $0.0\overline{6}$ more than 0.6.

6. a. $7/6 = 1.1\overline{6}$ b. $5/36 = 0.13\overline{8}$

7. a. 2.19   b. −1.45   c. 0.48

8. a. $-\dfrac{1}{9}$ b. $-\dfrac{69}{90} = -\dfrac{23}{30}$

9. a. $2.56 \cdot 10^7$ b. $7.81 \cdot 10^9$

10.

a. −0.003	b. −0.125
c. $-1\dfrac{28}{33}$	d. $8\dfrac{4}{27}$

11. a. 250   b. $81.\overline{81}$

12. Calculation: $\dfrac{1}{3} \cdot 12.75 = 12.75 \div 3 = 4.25$. For example: Suppose you need to pay 1/3 of

   the cost of medicine for the dog that costs $12.75. How much is your part? It is $4.25.

13. 15% of 3/4 can be calculated with decimals: $0.15 \cdot 0.75 = 0.1125$.

   Or, it can be calculated with fractions: $\dfrac{15}{100} \cdot \dfrac{3}{4} = \dfrac{3}{20} \cdot \dfrac{3}{4} = \dfrac{9}{80}$.

14. The number is −8.49. If two-thirds of the number is −5.66, then one-third of the number is half
   of the two-thirds, or −2.83. And the number itself is three times the one-third, or $3 \cdot (-2.83) = -8.49$.

15.

a. $\dfrac{\frac{2}{5}}{4}$	b. $\dfrac{\frac{9}{10}}{\frac{1}{6}}$
$\dfrac{2}{5} \div 4 = \dfrac{2}{5} \cdot \dfrac{1}{4} = \dfrac{2}{20} = \dfrac{1}{10}$	$\dfrac{9}{10} \div \dfrac{1}{6} = \dfrac{9}{10} \cdot \dfrac{6}{1} = \dfrac{54}{10} = 5\dfrac{2}{5}$

Mixed Review 5, p. 37

1. a. $3x + 9$ b. $3x + 9 = 57$ c. He bought 16 pairs of gloves.

2.

a. $2r - 5 = 10 - (-2)$	b. $2 \cdot 3 = 9 - 6y$
$2r - 5 = 12$ $\quad \mid +5$	$6 = 9 - 6y$ $\quad \mid -9$
$2r = 17$ $\quad \mid \div 2$	$-3 = -6y$ $\quad \mid \div (-6)$
$r = 8.5$	$1/2 = y$

3. a. It will take 1.28223 seconds. $t = d / v = (384{,}403 \text{ km}) / (299{,}792.458 \text{ km/s}) = 1.2822303889 \text{ s} \approx 1.28223 \text{ s}$.
   b. Please check the student's work. For example, according to Wikipedia, Mars made its closest approach to Earth in
   60,000 years on 27 August, 2003, at 09:51 Greenwich mean time at a distance of 55,758,006 km. At that point a
   radio signal would have taken 55,758,006 km / 299,792.458 km/sec = 185.98869 sec = 3 min 5.98869 sec, or about
   3 minutes and 6 seconds to arrive at Mars.

4.

a. $2 - x = -6$ $\quad\mid -2$ $-x = -8$ $x = 8$ Check: $2 - 8 \overset{?}{=} -6$ $-6 = -6$ ✓	b. $-10 - x = 7$ $\quad\mid +10$ $-x = 17$ $x = -17$ Check: $-10 - (-17) \overset{?}{=} 7$ $7 = 7$ ✓
c. $2x = -5$ $\quad\mid \div 2$ $x = -2.5$ Check: $2 \cdot (-2.5) \overset{?}{=} -5$ $-5 = -5$ ✓	d. $2 + (-11) = 8 + z$ $-9 = 8 + z$ $8 + z = -9$ $\quad\mid -8$ $z = -17$ Check: $2 + (-11) \overset{?}{=} 8 + (-17)$ $-9 = -9$ ✓

5. a. $-1/3$ b. $-4/5$ c. $1\,1/4$

6. a. $-4\,m + 2\,m = -2\,m$
   b. $-\$250 + (-\$500) = -\$750$

7.

a. $\dfrac{x}{-13} = 4$ $\quad\mid \cdot (-13)$ $x = 4 \cdot (-13)$ $x = -52$ Check: $\dfrac{-52}{-13} \overset{?}{=} 4$ $4 = 4$ ✓	b. $\dfrac{w}{-3} = -11 + 5$ $\dfrac{w}{-3} = -6$ $\quad\mid \cdot (-3)$ $w = -6 \cdot (-3)$ $w = 18$ Check: $\dfrac{18}{-3} \overset{?}{=} -11 + 5$ $-6 = -6$ ✓
c. $-31 = \dfrac{1}{6}x$ $\quad\mid \cdot 6$ $-186 = x$ $x = -186$ Check: $-31 \overset{?}{=} \dfrac{1}{6} \cdot (-186)$ $-31 = -31$ ✓	d. $1 = -5x$ $\quad\mid \div (-5)$ $-1/5 = x$ $x = -1/5$ Check: $1 \overset{?}{=} -5 \cdot (-1/5)$ $1 = 1$ ✓

1.

a. No. For example, when $n = 5$ and $m = 6$, we get $(-n - 1) - m = (-5 - 1) - 6 = $ **−12** $-n - (1 - m) = -5 - (1 - 6) = -5 - (-5) = $ **0**	b. Yes. The expressions are equal.

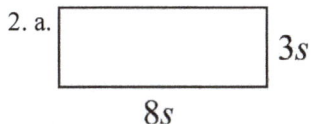

2. a.

         $3s$

     $8s$

   b. The area is $24s^2$.
   c. The perimeter is $22s$.

3.

a. $22r + 5$	b. $6p^2 + 8$	c. $42y^3$

4. a. $25n$
   b. $25n = 875$
        $n = 35$
     Ashley has 35 quarters.

5.

a. $20(5x + 3)$	b. $4(6s - t - 2)$

6. $(-67) + (-28) = -95$
   $(-67) + 28 = -39$
   $67 + (-28) = 39$
   $67 + 28 = 95$

7. a. Answers will vary. Please check the student's work. In the example below, the x values increase by 4 at each step, and the values of $3 - x$ decrease by 4 at each step.

x	$3 - x$
−9	$3 - (-9) = 12$
−5	$3 - (-5) = 8$
−1	$3 - (-1) = 4$
3	$3 - 3 = 0$
7	$3 - 7 = -4$
11	$3 - 11 = -8$

   b. When $x = 5$, the value of $3 - x$ is −2.

Mixed Review 6, cont.

8.

Balance	Equation	Operation to do to both sides
$xxxx \ominus$ $\oplus\oplus\oplus\oplus\oplus\oplus$	$4x - 1 = 7$	**+ 1**
↓		
$xxxx \oplus$ $\oplus\oplus\oplus\oplus\oplus\oplus$	$4x = 8$	
↓		
$xxxx$ $\oplus\oplus\oplus\oplus\oplus\oplus$	$4x = 8$	**÷ 4**
↓		
$xxxx$ $\oplus\oplus\oplus\oplus\oplus\oplus$	$x = 2$	

Equations and Inequalities Review, p. 41

1.

a.
$$
\begin{aligned}
1 - 3x &= 17 & &| -1 \\
-3x &= 16 & &| \div (-3) \\
x &= -5\,1/3
\end{aligned}
$$

Check: $1 - 3 \cdot (-5\,1/3) \overset{?}{=} 17$

$\quad\quad 1 - (-16) \overset{?}{=} 17$

$\quad\quad 1 + 16 \overset{?}{=} 17$

$\quad\quad\quad 17 = 17$ ✓

b.
$$
\begin{aligned}
29 &= -6 - 2y & &| + 2y \\
2y + 29 &= -6 & &| - 29 \\
2y &= -35 & &| \div 2 \\
y &= -17\,\tfrac{1}{2}
\end{aligned}
$$

Check: $\quad 29 \overset{?}{=} -6 - 2 \cdot (-17\,\tfrac{1}{2})$

$\quad\quad\quad 29 \overset{?}{=} -6 + 35$

$\quad\quad\quad 29 = 29$ ✓

c.
$$
\begin{aligned}
\frac{3x}{8} &= 42 & &| \cdot 8 \\
3x &= 336 & &| \div 3 \\
x &= 112
\end{aligned}
$$

Check: $\dfrac{3 \cdot 112}{8} \overset{?}{=} 42$

$\quad\quad 42 = 42$ ✓

d.
$$
\begin{aligned}
\frac{v - 2}{7} &= -13 & &| \cdot 7 \\
v - 2 &= -91 & &| + 2 \\
v &= -89
\end{aligned}
$$

Check: $\dfrac{-89 - 2}{7} \overset{?}{=} -13$

$\quad\quad \dfrac{-91}{7} \overset{?}{=} -13$

$\quad\quad -13 = -13$ ✓

1.

e. $\dfrac{w}{40} - 7 = 19$ $\quad$ **+ 7** $\dfrac{w}{40} = 26$ $\quad$ **· 40** $w = 1{,}040$ Check: $\dfrac{1{,}040}{40} - 7 \overset{?}{=} 19$ $26 - 7 \overset{?}{=} 19$ $19 = 19$ ✔	f. $\dfrac{s+8}{-3} = -1$ $\quad$ **· (−3)** $s + 8 = 3$ $\quad$ **− 8** $s = -5$ Check: $\dfrac{-5+8}{-3} \overset{?}{=} -1$ $\dfrac{3}{-3} \overset{?}{=} -1$ $-1 = -1$ ✔

2.

a.

Equation:	Logical thinking:
Let x be the normal price of one bottle of oil. $15x - 14 = 130$ $15x = 144$ $x = 144/15$ $x = \$9.60$	First, add the total cost and the discount to get the total cost of the oil bottles before the discount: \$130 + \$14 = \$144. Then we get the price of one bottle by dividing by 15: \$144 ÷ 15 = \$9.60.

b.

Equation:	Logical thinking:
Let x be the unknown number. $\dfrac{3}{7} x = 153$ $\quad$ **· 7** $3x = 1071$ $\quad$ **÷ 3** $x = 357$	Since three-sevenths of the number is 153, one-seventh of it is 153 ÷ 3 = 51. Then, the number itself is seven times that, or 7 · 51 = 357.

3. a. $300 + 18n$ or $18n + 300$

   b. $\quad 18n + 300 = 750 \quad$ **− 300**

   $\qquad\quad 18n = 450 \quad$ **÷ 18**

   $\qquad\qquad n = 25$

   Check:

   $18 \cdot 25 + 300 \overset{?}{=} 750$

   $\quad 450 + 300 = 750 \quad$ ✔

   He needs to sell 25 carpets in order to earn \$750 in a week.

4.

a.

$$2x + 6 + 3x = 9x - 11$$
$$5x + 6 = 9x - 11 \qquad | - 6$$
$$5x = 9x - 17 \qquad | - 9x$$
$$-4x = -17 \qquad | \div (-4)$$
$$x = 4\,\tfrac{1}{4}$$

Check:

$$2 \cdot (4\,\tfrac{1}{4}) + 6 + 3 \cdot (4\,\tfrac{1}{4}) \overset{?}{=} 9 \cdot (4\,\tfrac{1}{4}) - 11$$
$$8\,\tfrac{1}{2} + 6 + 12\,\tfrac{3}{4} \overset{?}{=} 38\,\tfrac{1}{4} - 11$$
$$27\,\tfrac{1}{4} = 27\,\tfrac{1}{4} \qquad \checkmark$$

b.

$$2(x + 6) = 9x - 11$$
$$2x + 12 = 9x - 11 \qquad | - 12$$
$$2x = 9x - 23 \qquad | - 9x$$
$$-7x = -23 \qquad | \div (-7)$$
$$x = 3\,\tfrac{2}{7}$$

Check:

$$2(3\,\tfrac{2}{7} + 6) \overset{?}{=} 9 \cdot (3\,\tfrac{2}{7}) - 11$$
$$2(9\,\tfrac{2}{7}) \overset{?}{=} 29\,\tfrac{4}{7} - 11$$
$$18\,\tfrac{4}{7} = 18\,\tfrac{4}{7} \qquad \checkmark$$

c.

$$6(5 - w) = 2(9 - w)$$
$$30 - 6w = 18 - 2w \qquad | + 2w$$
$$30 - 4w = 18 \qquad | - 30$$
$$-4w = -12 \qquad | \div (-4)$$
$$w = 3$$

Check:

$$6(5 - 3) \overset{?}{=} 2(9 - 3)$$
$$6(2) \overset{?}{=} 2(6)$$
$$12 = 12 \qquad \checkmark$$

d.

$$-10(4y + 7) = -9y$$
$$-40y - 70 = -9y \qquad | +70$$
$$-40y = -9y + 70 \qquad | + 9y$$
$$-31y = 70 \qquad | \div (-31)$$
$$y = -2\,\tfrac{8}{31}$$

Check:

$$-10(4(-2\,\tfrac{8}{31}) + 7) \overset{?}{=} -9(-2\,\tfrac{8}{31})$$
$$-10(-9\,\tfrac{1}{31} + 7) \overset{?}{=} 20\,\tfrac{10}{31}$$
$$-10(-2\,\tfrac{1}{31}) \overset{?}{=} 20\,\tfrac{10}{31}$$
$$20\,\tfrac{10}{31} = 20\,\tfrac{10}{31} \qquad \checkmark$$

5. a.

$$x + 2x + 7x + x = 180$$
$$11x = 180 \qquad | \div 11$$
$$x = 180/11$$
$$x = 16\,4/11 \approx 16.36$$

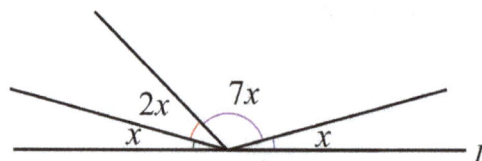

b. The measures of the four angles as fractions are: $x = 16\,4/11°$, $2x = 32\,8/11°$, $7x = 114\,6/11°$, and $x = 16\,4/11°$. However, angles aren't usually measured in elevenths of a degree, so it is more natural to give the answer in decimals. Therefore, the measures of the angles, rounded to two decimal digits, are 16.36°, 32.73°, 114.55°, and 16.36°.

6. One side of the rectangle is 12 ft, and the other sides is 7 ft + s. We can write the equation $12(7 + s) = 200$ for the total area.

$$12(7 + s) = 200$$
$$84 + 12s = 200 \qquad | - 84$$
$$12s = 116 \qquad | \div 12$$
$$s = 9\,2/3$$

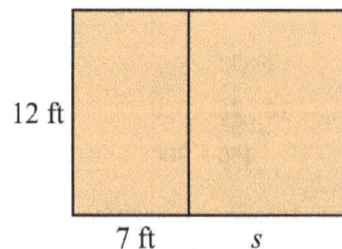

The unknown side measures 9 2/3 ft or 9 ft 8 in.

7.

a.	b.
$5x - 8 < 22$ $5x < 30$ $x < 6$	$x + 5 \geq -2$ $x \geq -7$

8. $-31 + 3(y + 5) = 23$

$-31 + 3y + 15 = 23$

$-16 + 3y = 23$ $\quad$ |+ 16

$3y = 39$ $\quad$ |÷ 3

$y = 13$

9. a. After the camera, clothes, and personal items she has 20 kg − 2.6 kg − 9 kg = 8.4 kg that she can use for the nuts. Ten bags of nuts weigh 8 kg. Eleven bags weigh 8.8 kg. So she can take only ten bags of nuts.

b. $0.8n + 2.6 + 9 \leq 20$

$0.8n + 11.6 \leq 20$ $\quad$ |− 11.6

$0.8n \leq 8.4$ $\quad$ |÷ 0.8

$n \leq 10.5$

She can take 10 bags of nuts.

10. a. $y = -2x - 1$

The slope is −2. Data in the table will vary.

x	2	1	0	−1	−2	−3	−4	−5
y	−5	−3	−1	1	3	5	7	9

b. The slope is 2 ½.

x	−3	−2	−1	0	1	2	3
y	−6 ½	−4	−1 ½	1	3 ½	6	8 ½

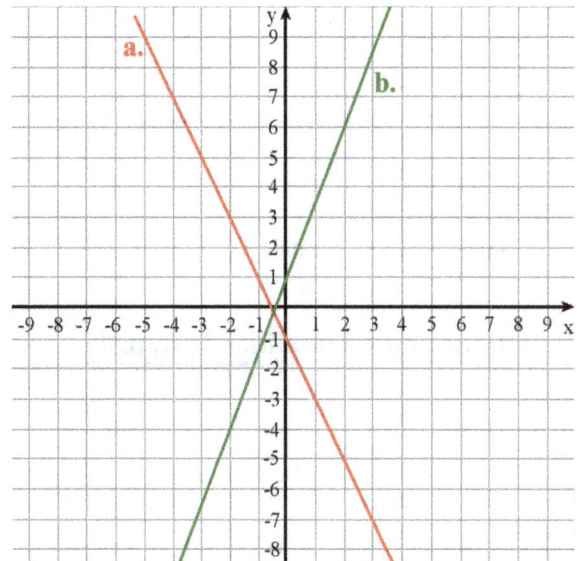

181

Equations and Inequalities Review, cont.

11. Answers will vary. Check the student's answer. The student's line should be parallel to the two purple example lines in the image on the right.

12. See the green line in the image on the right.

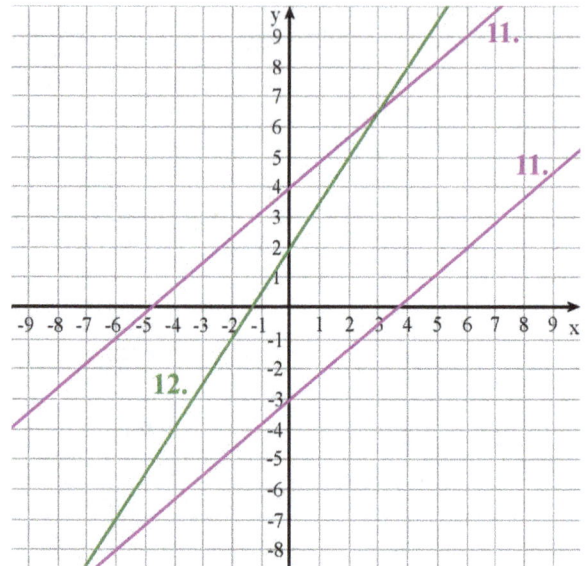

13. a. $d = 600t$. The distance (d) is in miles, and the time (t) is in hours.

b. Notice the line should go through the point (2.5, 1500) on the grid. (At 2 1/2 hours, the plane has flown 1,500 km).

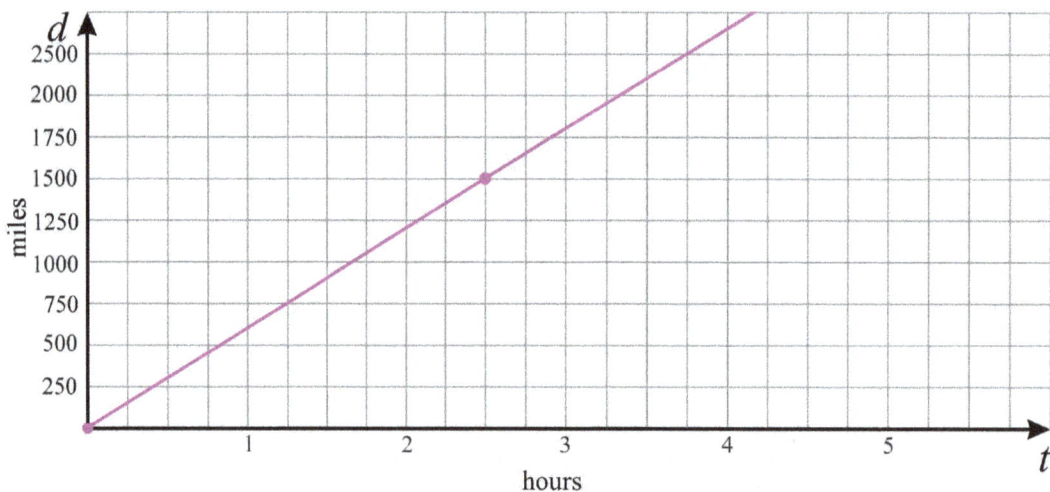

c. Since 1 hour 40 minutes = 1 2/3 hours, in that time the plane has flown $d = (1\ 2/3\ \text{h}) \cdot 600\ \text{mi/h} = 1{,}000$ miles.

Equations and Inequalities Test, p. 47

1.

a.			
-2	$=$	$6x + 5$	$\vert -5$
-7	$=$	$6x$	$\vert \div 6$
$-7/6$	$=$	x	
x	$=$	$-7/6$	

b.			
$6x + 2x - 1$	$=$	$-9x + 1$	
$8x - 1$	$=$	$-9x + 1$	$\vert +9x$
$17x - 1$	$=$	1	$\vert +1$
$17x$	$=$	2	$\vert \div 17$
x	$=$	$2/17$	

1.

c. $\dfrac{3x}{5} = 24$ $\quad\vert \cdot 5$ $3x = 120$ $\quad\vert \div 3$ $x = 40$	d. $\dfrac{y}{3} - 21 = -5$ $\quad\vert + 21$ $\dfrac{y}{3} = 16$ $\quad\vert \cdot 3$ $y = 48$

2. Let c be the cost of one cookie. We get the equation $24c + 3.25 = 6.85$. The cost of one cookie is $0.15.

$$24c + 3.25 = 6.85 \qquad \vert - 3.25$$
$$24c = 3.6 \qquad \vert \div 24$$
$$c = 0.15$$

3.

a. $\quad 3x + 5 < 68$ $3x < 63$ $\mathbf{x < 21}$	b. $\quad 10x - 17 \geq 103$ $10x \geq 120$ $\mathbf{x \geq 12}$

4. a. Let x be the number of layers of block. Then, the height of the shed, in inches, is $4 + 6 + 8x$ or $8x + 10$. Since the maximum height is 9.5 ft = 114 in, the inequality becomes $8x + 10 \leq 114$.

b. $\quad 8x + 10 \leq 114$
$$8x \leq 104$$
$$\mathbf{x \leq 13}$$

c.

Please note that in real life, there is also a limitation on the *minimum* height of the shed, so a builder can't choose to use fewer than about 10 or 11 rows.

5. a. $y = x + 4$

b. $y = -3x + 5$

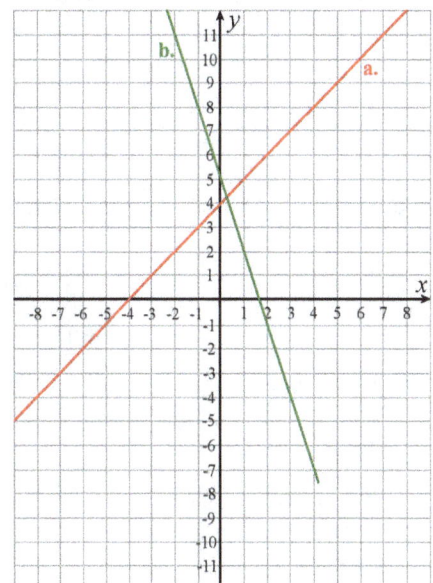

6. a. Answers will vary. Check the student's answer. The student's line
      should be parallel to the blue lines in the image on the right.

   b. See the image on the right. (The purple line has a slope of 1/2 and
      goes through the point (1, 3).)

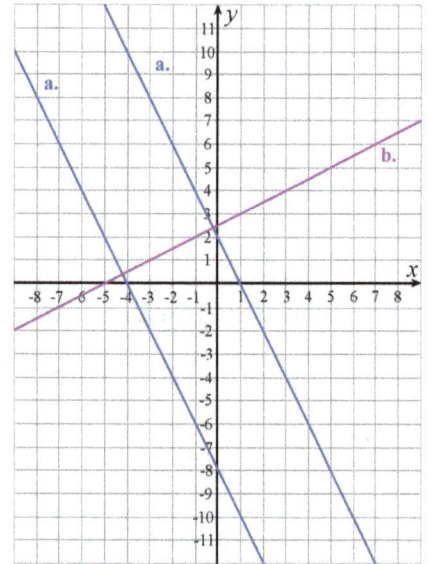

7. The slope is (change in y)/(change in x) = 10/1 = <u>10</u>.

8. a.

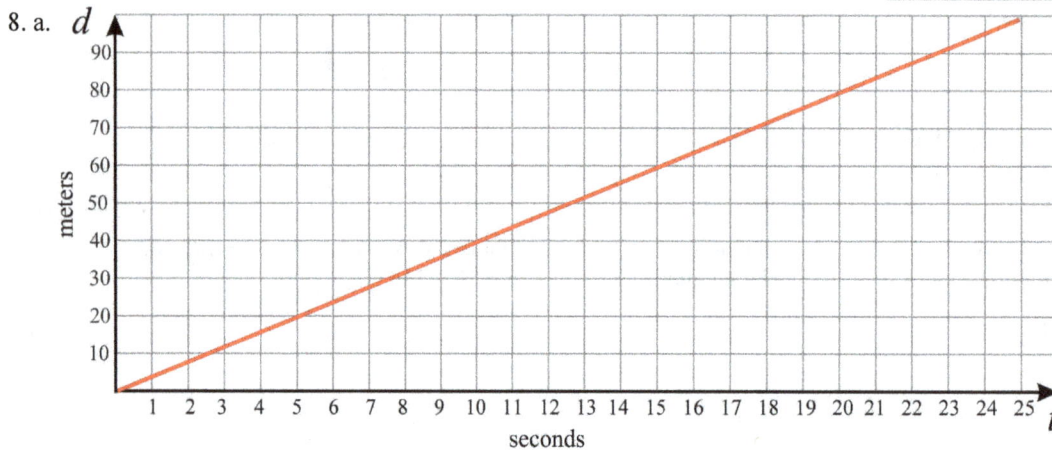

   b. Leah takes 25 seconds to run the 100 meters.
   c. $d = 4t$

Mixed Review 7, p. 51

1. a. **12**(6x − 5) = 72x − 60
   b. 12(**9y − 3x + 0.4**) = 108y − 36x + 4.8

2. $5x^3 + (-1/2)$ or $5x^3 − 1/2$

3. a. $p − 2/3p = 2600$
   b. Since 1/3 of the town's population is 2600, the total population used to be 3 · 2,600 = 7,800.

4. a. (−14) + 7 + (−8) + 2 = −13
   b. −3 + (−12) + 21 + (−19) + (−5) = −18

5. Check the student's answers for real-life situations.

   a. 1.4 · 119 = 166.6
      For example, an image that is 119 pixels wide is enlarged by 140%. How wide does it become?
      Answer: 167 pixels

   b. (9/10) · 14.30 = 12.87
      For example, an umbrella that costs $14.30 is on sale for 10% off. How much does it cost now?
      Answer: $12.87

6.

a. $-8-(-7)-(-12)$ $=-8+7+12=\underline{\textbf{11}}$	b. $63-(-11)+(-5)$ $=63+11+(-5)=\underline{\textbf{69}}$

7. a. $|x-8|$ or $|8-x|$ The latter is technically the distance between 8 and x, but naturally the distance between 8 and x is the same as the distance between x and 8.

   b. $|-52-8|=|-60|=60$

8.

a. $0.24 \div 0.03$	Decimal division: $0.24 \div 0.03$ $=2.4 \div 0.3$ $=24 \div 3 = \underline{\textbf{8}}$	Fraction division: $\dfrac{24}{100} \div \dfrac{3}{100}$ $\downarrow \quad \downarrow$ $\dfrac{24}{\cancel{100}} \cdot \dfrac{\cancel{100}}{3} = \dfrac{24}{3} = \underline{\textbf{8}}$
b. $7.1 \cdot 0.5$	Decimal multiplication: $\begin{array}{r} 7.1 \\ \cdot\ 0.5 \\ \hline 3.5\,5 \end{array}$	Fraction multiplication: $\dfrac{71}{10} \cdot \dfrac{\overset{1}{\cancel{5}}}{\underset{2}{\cancel{10}}}$ $\downarrow \quad \downarrow$ $\dfrac{71}{10} \cdot \dfrac{1}{2} = \dfrac{71}{20} = 3\,\dfrac{11}{20}$

9.

$\dfrac{5}{6} \cdot \dfrac{2}{3} \div \dfrac{4}{3}$ $\downarrow \quad \downarrow$ $\dfrac{5}{\underset{3}{\cancel{6}}} \cdot \dfrac{\overset{1}{\cancel{2}}}{\cancel{3}} \cdot \dfrac{\overset{1}{\cancel{3}}}{4}$ $\downarrow \quad \downarrow \quad \downarrow$ $\dfrac{5}{3} \cdot \dfrac{1}{1} \cdot \dfrac{1}{4} = \dfrac{5}{12}$	First, we change the division into a multiplication. Now we can simplify before multiplying.

10. a. \$1.65   b. -10.8 c. -150 m

11.

a. $0.5 \cdot \dfrac{11}{12}$ $= \dfrac{1}{2} \cdot \dfrac{11}{12} = \dfrac{11}{24}$ or $0.5 \cdot (11/12)$ $= 0.5 \cdot 11 \div 12 = 0.4583333\ldots$	b. $\dfrac{2}{5} \cdot (-0.8)$ $= \dfrac{2}{5} \cdot \left(-\dfrac{4}{5}\right) = -\dfrac{8}{25}$ or $(2/5) \cdot (-0.8)$ $= 0.4 \cdot (-0.8) = -0.32$	c. $-\dfrac{5}{6} \cdot 0.2$ $= -\dfrac{5}{6} \cdot \dfrac{1}{5} = -\dfrac{1}{6}$ or $(-5/6) \cdot 0.2$ $= -5 \div 6 \cdot 0.2 = -0.16666\ldots$

Mixed Review 7, cont.

12.

a. $2 + (-g) = 2 - g$ or $-g + 2$	b. $15 - (-r) = 15 + r$ or $r + 15$	c. $7x + (-2y) = 7x - 2y$

13. a. $1.13 \cdot 10^5$ b. $4.598 \cdot 10^7$

14.

a. $\dfrac{\frac{7}{8}}{9}$ $= \dfrac{7}{1} \div \dfrac{8}{9} = \dfrac{7}{1} \cdot \dfrac{9}{8}$ $= \dfrac{63}{8} = 7\dfrac{7}{8}$	b. $\dfrac{\frac{1}{2}}{\frac{1}{5}}$ $= \dfrac{1}{2} \div \dfrac{1}{5} = \dfrac{1}{2} \cdot \dfrac{5}{1}$ $= \dfrac{5}{2} = 2\dfrac{1}{2}$	c. $\dfrac{\frac{15}{21}}{\frac{2}{3}}$ $= \dfrac{15}{21} \div \dfrac{2}{3} = \dfrac{15}{21} \cdot \dfrac{3}{2}$ $= \dfrac{15}{14} = 1\dfrac{1}{14}$

Mixed Review 8, p. 54

1. a. **(iv)** $|x - 6|$ b. $|-23 - 6| = |-29| = 29$

2. (1) Substitute $a = -1$, $b = 1$, and $c = -1$ into the formula for the distributive property $a(b + c) = ab + ac$.
   $-1(1 + (-1)) = -1 \cdot 1 + (-1) \cdot (-1)$

   (2) The whole left side is zero because **1** + **(-1)** = 0.

   (3) So the right side must equal zero as well.

   (4) On the right side, $-1 \cdot 1$ equals **-1**. Therefore, $-1 \cdot (-1)$ must equal **1** so that the sum on the right side will equal zero.

   (5) Therefore, $-1 \cdot (-1)$ must equal **1**.

3.

a. $(-7) \cdot 2 \cdot (-2) = 28$	b. $10 \cdot (-4) \cdot 7 = -280$	c. $2 \cdot (-5) \cdot (-2) \cdot (-5) = -100$

4. Their average is $-3.7°C$:

$$\frac{-8°C + (-11°C) + 2°C + 0°C + (-3°C) + (-5°C) + (-1°C)}{7} = -3 \, 5/7°C \approx -3.7°C$$

5.

a. $0.3 \cdot 2.5 = 0.75$	b. $-0.002 \cdot 0.008 = -0.000016$	c. $-0.9 \cdot 50 = -45$
d. $0.8^2 = 0.64$	e. $-4 \cdot 0.05 \cdot (-20) = 4$	f. $(-0.3)^2 = 0.09$

6. a. -0.00061 b. 9807.2 c. 0.055191

7. Alex took <u>26.4 minutes</u> to commute that day.
   From the formula $d = vt$ we get that $t = d/v$. Going to work took Alex $t_1 = 12$ km/(60 km/h) = 1/5 h = 12 minutes, and coming back took him $t_2 = 12$ km/(50 km/h) = 6/25 h = 14.4 minutes. In total, the commuting took him 12 min + 14.4 min = 26.4 minutes.

8. a. $2.208\overline{3}$ b. 0.292857 c. 280

9.

a.	b.
$\dfrac{3}{5} + \left(-\dfrac{2}{3}\right)$	$-\dfrac{1}{2} + \left(-\dfrac{6}{9}\right)$
$\dfrac{9}{15} + \left(-\dfrac{10}{15}\right)$	$= -\dfrac{9}{18} + \left(-\dfrac{12}{18}\right)$
$= \dfrac{9-10}{15} = -\dfrac{1}{15}$	$= \dfrac{-9-12}{18} = -\dfrac{21}{18} = -1\dfrac{3}{18} = -1\dfrac{1}{6}$

Ratios and Proportions Review, p. 56

1.

a. 41 km per hour	b. $\dfrac{3\text{ g}}{800\text{ ml}}$	c. $1:3$

2.

Miles	58	116	174	232	290	348	580	1,160
Hours	1	2	3	4	5	6	10	20

3. $20\text{ g} : 1{,}200\text{ g} = 1:60$

4. Susan jogs at a rate of $\dfrac{1\frac{1}{2}\text{ mi}}{1/3\text{ h}} = \dfrac{3/2\text{ mi}}{1/3\text{ h}} = (3/2) \cdot (3/1)\text{ mi/h} = 9/2\text{ mi/h} = 4\frac{1}{2}\text{ mi/h}.$

5.

a.	b.
$\dfrac{16}{17} = \dfrac{109}{T}$	$\dfrac{1.5}{2.8} = \dfrac{M}{5}$
$16T = 17 \cdot 109$	$2.8M = 1.5 \cdot 5$
$16T = 1853$	$2.8M = 7.5$
$\dfrac{16T}{16} = \dfrac{1853}{16}$	$\dfrac{2.8M}{2.8} = \dfrac{7.5}{2.8}$
$T \approx 115.81$	$M \approx 2.68$

6.

$$\frac{\$19}{12\text{ kg}} = \frac{p}{5\text{ kg}}$$

$$12\text{ kg} \cdot p = \$19 \cdot 5\text{ kg}$$

$$\frac{12\text{ kg} \cdot p}{12\text{ kg}} = \frac{\$19 \cdot 5\text{ kg}}{12\text{ kg}}$$

$$p = \$7.91\overline{6} \approx \underline{\$7.92}$$

7. Since $8:10 = 4:5 = 20:25$, Gary can expect to make 20 baskets when he practices 25 shots.

8.

a. $\dfrac{2\ 1/2\ \text{pages}}{1\ 1/4\ \text{h}} = \dfrac{5/2\ \text{pages}}{5/4\ \text{h}} = (5/2) \cdot (4/5)\ \text{pages/h} = 2\ \text{pages/hour}$
Alex solved problems at a rate of 2 pages per hour.
b. $\dfrac{2/3\ \text{room}}{3/4\ \text{h}} = (2/3) \cdot (4/3)\ \text{room per hour} = 8/9\ \text{room/hour}$
Noah painted at a rate of 8/9 room in one hour.

9. A car is traveling at a constant speed of 75 km per hour.

   a. $d = 75t$. See the plot below.
   b. The unit rate is 75 km/h.
   c. See the graph below.

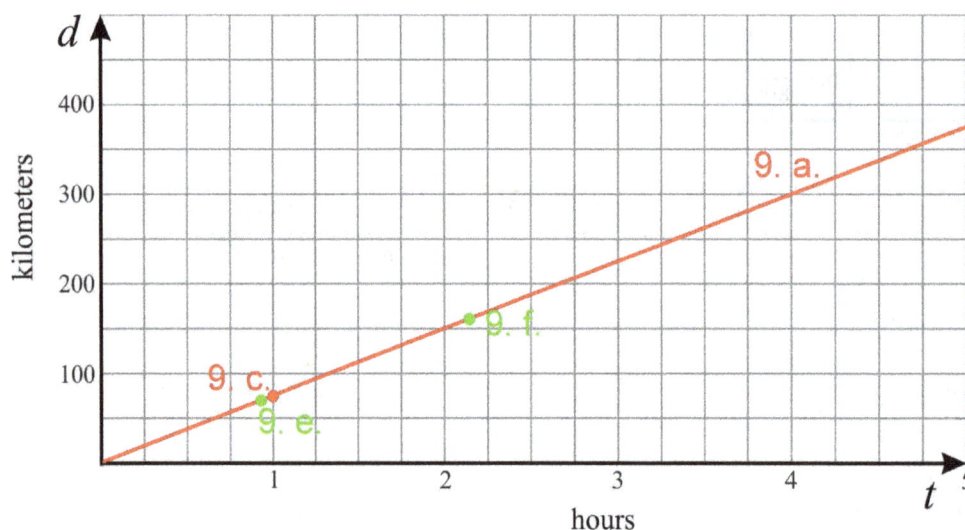

   d. The point (0, 0) mean the car has gone zero km in zero hours.

   e. Since 55 minutes is 55/60 = 11/12 hour, the car can travel $d = 75 \cdot (11/12) = 68.75$ km ≈ 69 km in 55 minutes. See the grid above for the point (11/12 h, 69 km).

   f. From the equation $75t = 160$ we get $t = 160/75 = 2.13333...$ hours ≈ 2 hours 8 minutes. See the grid above for the point (2.1 h, 160 km).

10. a. The quantities not in proportion. For example, 1 G for $10 gives us the unit rate of $10 per gigabyte, whereas paying $30 for 10G would give the unit rate of $3 per gigabyte. If they were in proportion, the unit rate would be the same no matter which two values you use to calculate it. You can also see it from the fact that whenever the bandwidth increases by 5G, the price increases sometimes $7, sometimes $6 so it does not always increase by the same amount.

    b. Does not apply.

11. It would cost $9,369 / 12 \cdot 5 \approx$ $3904 to drive the car for five months.

12. a. 8 m : 10 m = x : 6 m, from which x = 4.8 meters. Or you can reason that the sides of the smaller triangle are 0.8 of the sides of the bigger, so the unknown side is 0.8 · 6 m = 4.8 m.

    b. 8 in : 5.6 in = 13 in : x, from which x = 9.1 inches.

13. The true dimensions of the room are 2 in · 6 ft/1 in = 12 ft and 2 ¾ in. · 6 ft/1 in = 16 ½ ft.

14. a. The unit rate is 6 mi/gal or 6 mpg (miles per gallon).

   b. $M = 6f$

   c. Answers may vary because the scaling on the axes may vary. Check the student's plot. For example:

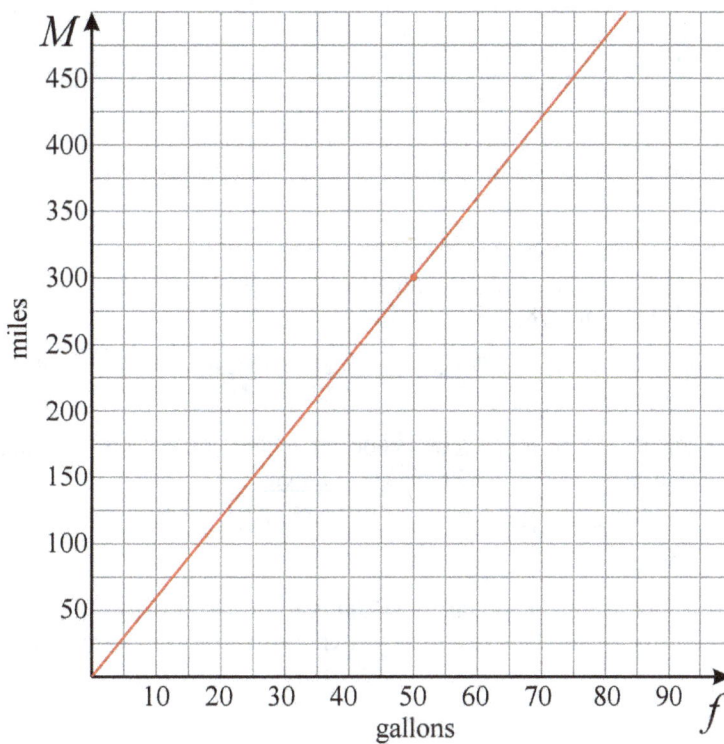

Ratios and Proportions Test, p. 61

1. Chloe's speed = (20 km) / (1 1/2 h) = (20 km) / (3/2 h) = 20 · 2/3 km/h = 40/3 km/h = 13 1/3 km/h

2. a. (1/3 envelope) / (2/3 cups) = (1/3) · (3/2) envelope/cup = 1/2 envelope/cup

   b. Mason is using 1/2 envelope per a cup of water.

3. Proportions vary as there are several different ways to write the proportion correctly. Here are four of the correct ways. Besides these four, you will get four more by switching the right and left sides of these four equations.

$\dfrac{52 \text{ kg}}{\$169} = \dfrac{21 \text{ kg}}{x}$	$\dfrac{\$169}{52 \text{ kg}} = \dfrac{x}{21 \text{ kg}}$	$\dfrac{21 \text{ kg}}{52 \text{ kg}} = \dfrac{x}{\$169}$	$\dfrac{52 \text{ kg}}{21 \text{ kg}} = \dfrac{\$169}{x}$

The key point is that in each of the correct ways, x ends up being multiplied by 52 kg in the cross-multiplication. If x ends up being multiplied by 21 kg or $169 in the cross-multiplication, the proportion is set up incorrectly.

Here is the solution process for one of the proportions above. Each of the others has the same final solution, $x = \$68.25$.

$$\frac{\$169}{52 \text{ kg}} = \frac{x}{21 \text{ kg}}$$

$$52 \text{ kg} \cdot x \qquad \$169 \cdot 21 \text{ kg}$$

$$x = \frac{\$169 \cdot 21 \text{ kg}}{52 \text{ kg}}$$

$$x = \$68.25$$

4.

$$\frac{4.3}{S} = \frac{7.9}{12}$$

$$7.9S = 4.3 \cdot 12$$

$$7.9S = 51.6$$

$$S = \frac{51.6}{7.9}$$

$$S \approx 6.5$$

5. The student can set up a proportion, such as $x/84$ m $= 56/48$, or use logical reasoning to solve the problem. The final solution is $x = (56/48) \cdot 84$ m $= \underline{98\ m}$.

6. The student can set up a proportion, such as $16/9 = x/63$ cm, or use logical reasoning to solve the problem. The width is 63 cm / 9 · 16 = 112 cm.

7. a. The scale 1:45,000 means that 1 cm corresponds to 45,000 cm, which equals 450 m. So we can rewrite the scale as 1 cm : 450 m. Then, 850 m corresponds to 850/450 cm ≈ <u>1.9 cm</u>.

   Another way to solve this is to first divide 850 m ÷ 45,000 = 0.01$\overline{8}$ m, and then convert 0.01$\overline{8}$ m into centimeters: 0.01$\overline{8}$ m = 1.$\overline{8}$ cm ≈ 1.9 cm.

   b. 5.4 cm on the map corresponds to 5.4 cm · 45,000 = 243,000 cm = 2,430 m = <u>2.43 km</u> in reality.
   Another way to solve this is to use the converted scale of 1 cm : 450 m to get 5.4 cm · 450 m = 2,430 m = 2.43 km.

8. a. The unit rate is 5 m/s or 5 meters per 1 second.

   b.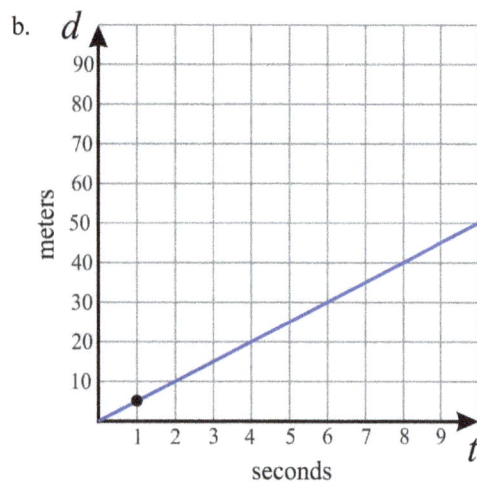

   c. $d = 5t$

9. a. The unit rate is $15 per hour.

   b. The graphs will vary because the scaling on the grid can vary. Check the student's graph. The point corresponding to $t = 10$ hours should fit on the graph. For example:

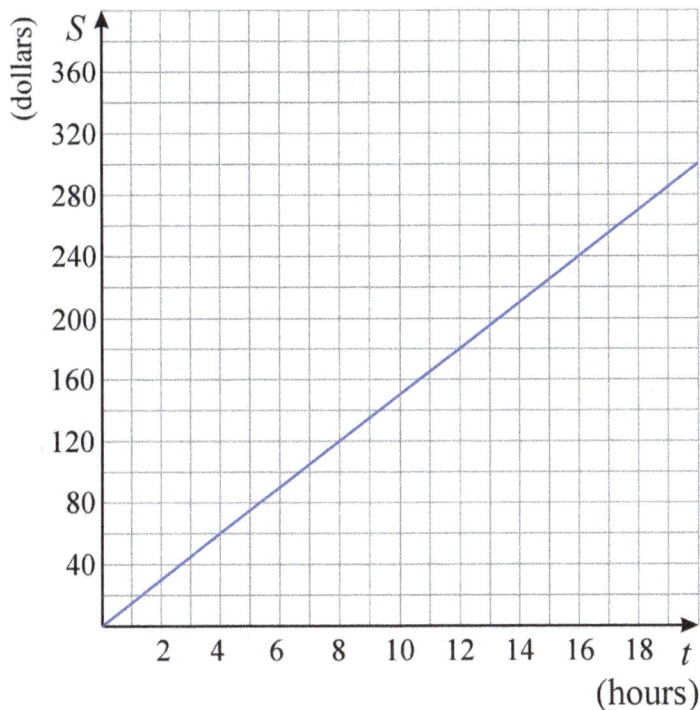

Mixed Review 9, p. 64

1. a. Yellowknife, NT   b. The difference is $21.3°C − (−21.6°C) = 42.9°C$.
   c. Vancouver, BC   d. The difference is $22.1°C − 6.8°C = 15.3°C$.

2. a. $2,089,000 = 2.089 \cdot 10^6$ b. $394,410,000 = 3.9441 \cdot 10^8$

3. b. $−1/6$ c. $−18/5$ or $−3\ 3/5$ d. $2/9$ e. $−6/7$ f. $8/7$ or $1\ 1/7$

4. a. $−1$ b. 10 c. 3 d. 6 e. 45 f. 48

5.

a.			b.		
$11 − 5x$	$=$	$−6$		$6(y + 2)$	$=$	$−16$
$− 5x$	$=$	$−17$		$6y + 12$	$=$	$−16$
$5x$	$=$	17		$6y$	$=$	$−28$
x	$=$	$17/5 = 3\ 2/5$		y	$=$	$−28/6 = −14/3 = −4\ 2/3$

a. Check: $11 − 5(17/5) \overset{?}{=} −6$
          $11 − 17 = −6$ ✓

b. Check: $6(−4\ 2/3 + 2) \overset{?}{=} −16$
          $6(−2\ 2/3) \overset{?}{=} −16$
          $6(−8/3) \overset{?}{=} −16$
          $−16 = −16$ ✓

5.

c. $\dfrac{2x}{5} = 30$	d. $\dfrac{s-12}{5} = -1$
$2x = 150$	$s - 12 = -5$
$x = 75$	$s = 7$

Check: $\dfrac{2(75)}{5} \overset{?}{=} 30$

$\dfrac{150}{5} = 30$ ✓

Check: $\dfrac{7-12}{5} \overset{?}{=} -1$

$\dfrac{-5}{5} = -1$ ✓

6. See the image on the right. The slope is $-7/3$ or $-2\,1/3$.

7. a. $T \le 42°C$ b. $f \ge 3$ C   b. $C < \$3,000$

8. a. In three months, one worker costs him $3 \cdot \$2,050 = \$6,150$.
Since $\$40,000 \div \$6,150 \approx 6.50406$, he can hire six workers.

b. The question asks for how many workers he can hire, and that gives us
our unknown: let n be the number of workers the farmer hires. Then, we
can write the inequality $n \cdot 3 \cdot \$2,050 < \$40,000$. Here's its solution:

$$n \cdot 3 \cdot \$2,050 \; < \; \$40,000$$
$$\$6,150n \; < \; \$40,000$$
$$n \; < \; \$40,000/\$6,150$$
$$n \; < \; 6.50406$$

So, he can hire six workers.

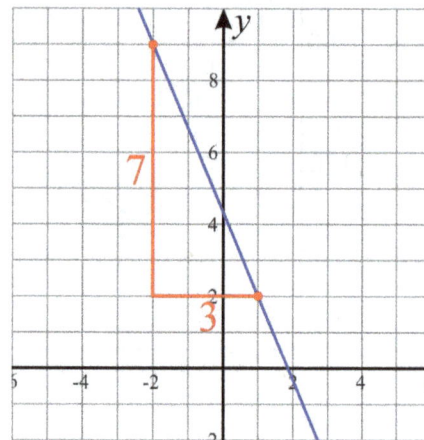

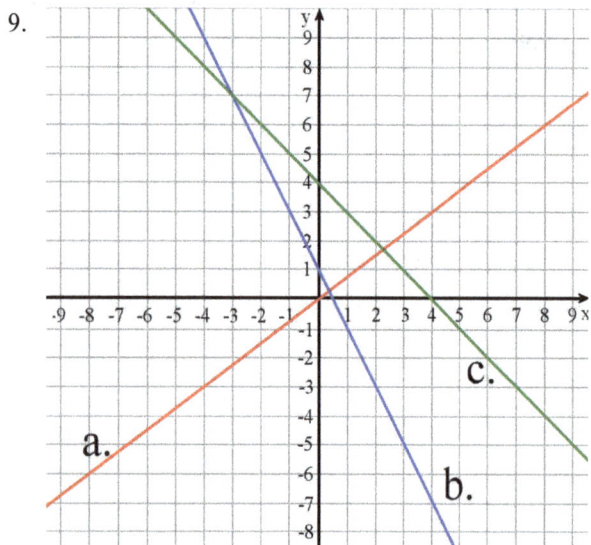

9.

10. We can check if the point $(2, -3)$ fulfills the equation $y = -2x - 2$:

$$-3 \overset{?}{=} -2(2) - 3$$

$$-3 \overset{?}{=} -4 - 3$$

$$-3 \ne -7$$

The point is not on the line. Another way to find this out is to plot the line and the point in a coordinate grid, and
notice that the point is not on the line.

1. a. $2t + 4z - 9$ (or $4z + 2t - 9$ but listing the terms in alphabetical order of the variable is preferred)

   b. $-42x^3$

   c. $24s^2t$

2. a. \$120   b. 4.6 kg   c. 0.008 or 8/1000

3.

<table>
<tr><td>

a.

$$2x - 6 = 9x - 8$$
$$-6 + 8 = 9x - 2x$$
$$2 = 7x$$
$$2/7 = x$$

Check:

$$2(2/7) - 6 \stackrel{?}{=} 9(2/7) - 8$$
$$4/7 - 6 \stackrel{?}{=} 18/7 - 8$$
$$4/7 - 42/7 \stackrel{?}{=} 18/7 - 56/7$$
$$-38/7 = -38/7 \checkmark$$

</td><td>

b.

$$3(x - 6) = -9x$$
$$3x - 18 = -9x$$
$$3x + 9x = 18$$
$$12x = 18$$
$$x = 18/12 = 1\tfrac{1}{2}$$

Check:

$$3(1\tfrac{1}{2} - 6) \stackrel{?}{=} -9(1\tfrac{1}{2})$$
$$3(-4\tfrac{1}{2}) \stackrel{?}{=} -13\tfrac{1}{2}$$
$$-13\tfrac{1}{2} = -13\tfrac{1}{2} \checkmark$$

</td></tr>
<tr><td>

c.

$$8x = -\frac{3}{4}$$
$$x = -\frac{3}{4} \div 8$$
$$x = -\frac{3}{32}$$

Check:

$$8(-3/32) \stackrel{?}{=} -3/4$$
$$-24/32 = -3/4 \checkmark$$

</td><td>

d.

$$1\frac{1}{6} + v = \frac{2}{9}$$
$$v = \frac{2}{9} - 1\frac{1}{6}$$
$$v = \frac{4}{18} - \frac{21}{18}$$
$$v = -\frac{17}{18}$$

Check:

$$1\,1/6 - 17/18 \stackrel{?}{=} 2/9$$
$$21/18 - 17/18 \stackrel{?}{=} 2/9$$
$$4/18 = 2/9 \checkmark$$

</td></tr>
</table>

4. a. 3

   b. 1/3

   c. −3/4

5. a. $4.9 \cdot 10^8$

   b. $6.238 \cdot 10^9$

6. a. 208,000,000   b. 1,293,000

7. See the image on the right.

8. Her average speed for the entire trip was <u>14 km/h</u>.
   The entire trip took Cynthia 24 minutes, and she bicycled 5.6 km.
   For calculating the average speed, we need the time in hours:
   24 minutes = 24/60 hours = 2/5 hours.

   Now, we can use the formula $d = vt$ and write 5.6 km = $v \cdot$ (2/5 h),
   from which v = 5.6 km/(2/5 h) = 5.6 $\cdot$ 5/2 km/h = 14 km/h.

9. a. The maximum width is approximately <u>8.89 ft or about</u>
   <u>8 ft 10 inches</u>.

   To solve the problem, first calculate the maximum area
   for the driveway: A = $1,600/($4.50/sq. ft) ≈ 355.556 sq. ft.

   Then, the maximum width is 355.556 sq. ft./(40 ft) ≈ 8.89 ft.

   b. Let w be the width of the driveway. The area of the driveway
   is then 40w. The cost of that is 4.50 $\cdot$ 40w. We can write the
   inequality 4.50 $\cdot$ 40w ≤ 1,600. Solution:

   $$4.50 \cdot 40w \le 1{,}600$$
   $$180w \le 1{,}600$$
   $$w \le 8.89$$

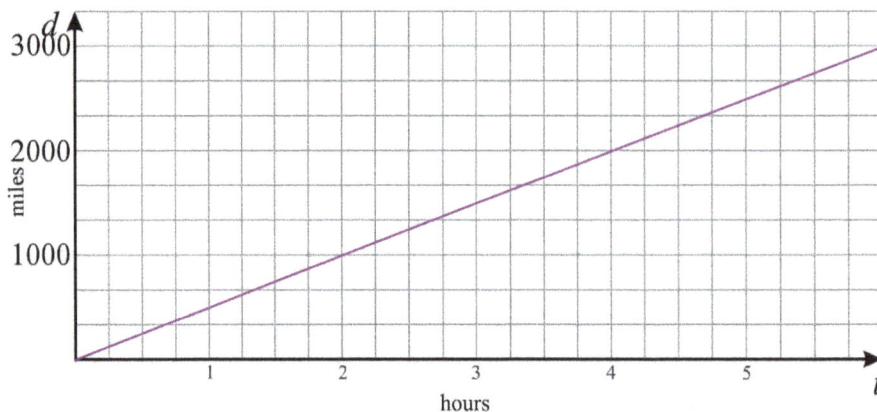

10. a. $d = 500t$

   b. Answers will vary because the scaling on the d-axis can vary.
   Check the student's answer. For example:

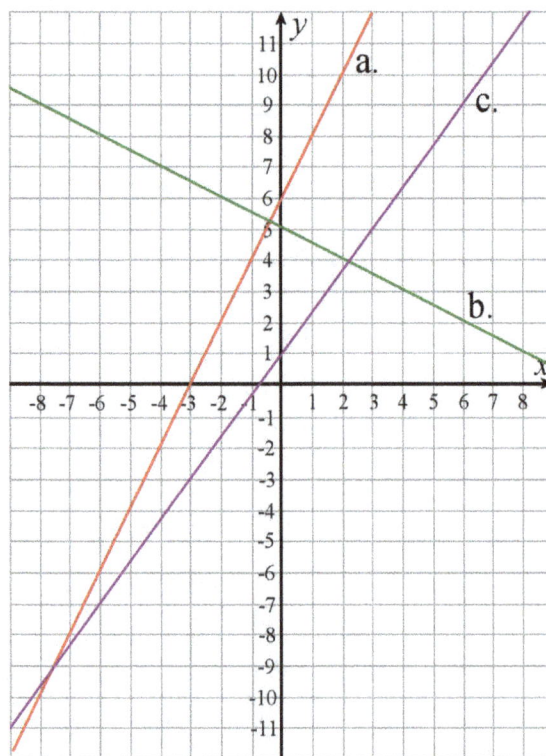

   c. We can substitute 3,600 in place of d in the equation $d = 500t$ to get the equation 3,600 = 500t. Its solution is
   t = 3,600/500 = 7.2 hours = 7 hours 12 minutes

Percent Review, p. 70

1. a. 0.092 $\cdot$ $150 = $13.80.   b. 0.458 $\cdot$ 16 m = 7.328 m.   c. 0.006 $\cdot$ 700 mi = 4.2 mi.

2. a. $9 $\cdot$ 0.80 = $7.20. New price: $7.20.   b. $6 $\cdot$ 0.75 = $4.50. New price: $4.50.   c. $90 $\cdot$ 0.70 = $63. New price: $63.

3. 0.82p = $23.37
   0.82p/0.82 = $23.37/0.82
       p = $28.50

4. The price of the computer with tax is $459 $\cdot$ 1.07 = $491.13.
   Andy's share of it is $491.13 $\cdot$ 0.4 = <u>$196.45</u>.   Jack's share of it is $491.13 $\cdot$ 0.6 = <u>$294.68</u>.

5. The discount percentage is $27/$180 = 3/20 = 0.15 = 15%.

6. a. The scale factor between the triangles is 21 cm / 7 cm = 1/3. The unknown side is therefore 12 cm/3 = 4 cm.
The area of the larger triangle is 12 cm · 21 cm/2 = 126 cm^2.
The area of the smaller triangle is 4 cm · 7 cm/2 = 14 cm^2.
The area of the smaller triangle is 14/126 = 11.1% of the area of the larger triangle.

   b. If you take the ratio from the larger to the smaller triangle, the sides of the triangles are in the ratio of 3:1.
Going from the smaller one to the larger one, the ratio is 1:3.

   c. If you take the ratio from the larger to the smaller triangle, the areas are in a ratio of 126:14 = 63:7 = 9:1.
Going from the smaller one to the larger one, the ratio is 1:9.

7. Originally, the area of the wall painting would have been 5 m · 3 m = 15 m^2.
After scaling, the sides are 5 m · 1.2 = 6 m and 3 m · 1.2 = 3.6 m, so the enlarged area is 6 m · 3.6 m = 21.6 m^2.
The difference in area is 21.6 m^2 − 15 m^2 = 6.6 m^2, so the percentage increase is (*difference in area*)/(*original area*)
= 6.6 m^2/15 m^2 = 6.6/15 = 0.44 = 44%. An easier way to figure this is just to realize that, regardless of the actual
dimensions, since the scaling of each side is 1.2, the scaling of the area is just 1.2 · 1.2 = 1.44, so the increase in
area is 44%.

8. The difference in their times is 200 sec − 120 sec = 80 sec. Their average time was ½(200 sec + 120 sec) = 160 sec.

   a. (*Difference*)/(*The Old Gray Mare's time*) = 80/200 = 4/10 = 40%. Old Paint was 40% quicker than the Old
Gray Mare.

   b. (*Difference*)/(*Old Paint's time*) = 80/120 = 2/3 = 66.7%. The Old Gray Mare was 66.7% slower than Old Paint.

   c. (*Difference*)/(*Average time*) = 80/160 = 1/2 = 50%. The relative difference between the two horses was 50%.

9. Assuming none of the interest was added to the principal during the time of the loan, he would owe the total amount
of interest at the end of two years. For the first year, the interest is $4,000 · 0.078 = $312, and for the second, it is
$2,000 · 0.078 = $156. The total interest is $312 + $156 = $468.

Percent Test, p. 72

1. a. New price: 0.88 · $110 = $96.80
   b. New price: 1.024 · $5,000 = $5,120
   c. Discount percent: ($90 − $59)/$90 ≈ 34.4%

2. The base price of the purchase is 3 · $10 + 2 · $20 = $70. With a 5% discount, this becomes $70 − $3.50 = $66.50.
Then adding the sales tax, the final price is 1.062 · $66.50 = $70.623 ≈ $70.62.

3. Let n be the number of students the college had last year. The 6.6% increase means that n increased by 0.066n
becomes 1,210. As an equation, this is n + 0.066n = 1,210, from which n = 1,210/1.066 ≈ 1,135 students.

4. a. It lost 2.3 kg/25 kg = 0.092 = 9.2% of its body weight.

   b. Mary would weigh (1 − 0.092) · 58 kg = 0.908 · 58 kg = 52.664 kg ≈ 52.7 kg.

5. The original area is 5 m · 6.5 m = 32.5 m^2. The area after the enlarging is 7.2 m · 10 m = 72 m^2.

   The percent increase is (72 m^2 − 32.5 m^2)/32.5 m^2 = 1.2$\overline{153846}$ ≈ 122%.

6. a. (10,779,264 − 10,320,257)/10,320,257 = 0.0444763 ≈ 4.4% more males than females.

   b. (11,282,003 − 10,827,017)/10,827,017 = 0.0420232 ≈ 4.2% more females than males.

7. The pizza in PizzaTown is ($15.99 − $12.99)/$12.99 = 0.23094688 ≈ 23.1% more expensive than the pizza in
Tony's Pizzeria.

8. She will have $2,500 · 0.044 · 3 + $2,500 = $2,830 after three years.

9. The actual interest he paid was $65,800 − $35,000 = $30,800. The interest he accumulated in one year was
$30,800 ÷ 10 = $3,080. Then, the yearly interest rate was $3,080/$35,000 = 0.088 = 8.8%.

1. a. $3(-4)/(-4+7) = -12/3 = \underline{-4}$

   b. $(1-7)/(1+7) = -6/8 = \underline{-3/4}$

2. a.

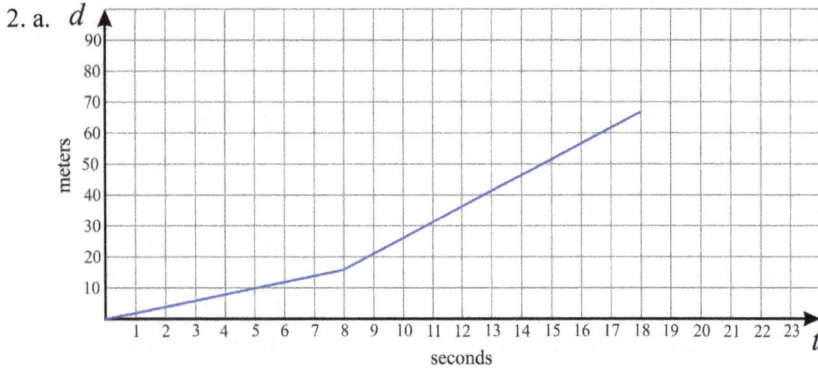

   b. He runs a total of $(8 \text{ s} \cdot 2 \text{ m/s}) + (10 \text{ s} \cdot 5 \text{ m/s}) = 16 \text{ m} + 50 \text{ m} = \underline{66 \text{ m}}$.

3.

a. $\dfrac{6x}{7} = -1.2$ $6x = -8.4$ $x = -1.4$	b. $\begin{aligned} 6x - 7 &= -1.2 \\ 6x &= 7 - 1.2 \\ 6x &= 5.8 \\ x &= 5.8/6 \approx 0.9667 \end{aligned}$

4. a. Yes, the expressions are equal.
   b. No, the expressions are not equal. For example, when $x = 1$ and $y = 0$, we get $x - 2y = 1 - 0 = 1$ whereas $y - 2x = 0 - 2 = -2$.

5. a. $|a - b| = |-5 - 6| = |-11| = 11$. This is correct because the numbers -5 and 6 are 11 units apart on the number line.

   b. $|a - b| = |-2 - (-11)| = |-2 + 11| = |9| = 9$. This is correct because the numbers -2 and -11 are 9 units apart on the number line.

6. a. $|x - 7|$ or $|7 - x|$

   b. $|x - 7| = |-3 - 7| = |-10| = 10$

7. a. $-(-2)$ which simplifies to 2
   b. $|-80|$ which simplifies to 80
   c. $-(6 + 7)$, which simplifies to -13
   d. $|-4 + 5|$, which simplifies to 1

8.

a. $\dfrac{14 \text{ mi}}{0.59 \text{ gal}} = \dfrac{100 \text{ mi}}{V}$ $14 \text{ mi} \cdot V = 100 \text{ mi} \cdot 0.59 \text{ gal}$ $V = \dfrac{100 \text{ mi} \cdot 0.59 \text{ gal}}{14 \text{ mi}}$ $V \approx 4.21 \text{ gal}$	b. $\dfrac{P}{2000 \text{ lb}} = \dfrac{\$4.05}{3 \text{ lb}}$ $P \cdot 3 \text{ lb} = \$4.05 \cdot 2000 \text{ lb}$ $P = \dfrac{\$4.05 \cdot 2000 \text{ lb}}{3 \text{ lb}}$ $P = \$2,700$

9. The cost is $(15/21) \cdot \$3.14 = \underline{\$2.24}$.

10. There are several ways to solve this problem:

    (1) The unknown side is $50/60 = 5/6$ of the longer side so it is $(5/6) \cdot 45$ in = <u>37.5 in.</u>

    (2) The sides of the smaller trapezoid are $45/60 = 3/4$ of the sides of the bigger trapezoid, so the unknown size is $(3/4) \cdot 50$ in = <u>37.5 in.</u>

    (3) We can write the proportion 50 in/60 in = x/45 in and solve it to get $x = 37.5$ in.

    (4) We can write the proportion 50 in/x = 60 in/45 in and solve it to get $x = 37.5$ in.

    (3) We can write the proportion 60 in/50 in = 45 in/x and solve it to get $x = 37.5$ in.

    (5) We can write the proportion x/50 in = 45 in/60 in and solve it to get $x = 37.5$ in.

11.

$0.6 \cdot 0.7$	Decimal multiplication:	Fraction multiplication:
	$0.6 \cdot 0.7 = 0.42$	$\dfrac{6}{10} \cdot \dfrac{7}{10} = \dfrac{42}{100} = \dfrac{21}{50}$

12.

$0.24 \div 0.5$	Decimal division:	Fraction division:
	$0.24 \div 0.5$ has the same answer as $2.4 \div 5$.	$\dfrac{24}{100} \div \dfrac{5}{10} = \dfrac{24}{100} \cdot \dfrac{10}{5} = \dfrac{6}{25} \cdot \dfrac{2}{1} = \dfrac{12}{25}$

$$\begin{array}{r} 0.4\,8 \\ 5)\overline{2.4\,0} \\ \underline{-2\,0} \\ 4\,0 \\ \underline{-4\,0} \\ 0 \end{array}$$

The fraction $12/25$ is equivalent to $48/100 = 0.48$.

Mixed Review 12, p. 77

1.

a.	b.
$3y + 7 \; < \; 56$	$-5 + 6z \; \geq \; 175$
$3y \; < \; 49$	$6z \; \geq \; 180$
$y \; < \; 49/3 = 16\,1/3$	$z \; \geq \; 30$

2. Let n be the number sales in a week. I will earn $180 + 45n$, and if I want to earn at least \$500, the inequality is:

$$180 + 45n \; \geq \; 500$$
$$45n \; \geq \; 320$$
$$n \; \geq \; 7.\overline{1}$$

Since I cannot sell fractional parts of a painting, I need to sell <u>at least 8 paintings</u> in order to earn at least \$500 in a week.

3.

Jim can swim 30 laps in a pool in 26 minutes. How many laps could he swim in 45 minutes? **Matt's Answer:** He could swim 30 laps. **Solution:** $\dfrac{30 \text{ laps}}{26 \text{ min}} = \dfrac{L}{45 \text{ min}}$ $26L = 30 \cdot 26$ $26L = 780$ $\dfrac{26L}{26} = \dfrac{780}{26}$ $L = 30$	Matt wrote the proportion correctly, but did not do the cross-multiplying correctly. He multiplied 30 by 26 whereas he should have multiplied 30 by 45. **Correct answer:** He could swim <u>almost 52 laps</u>. **Correct solution:** $\dfrac{30 \text{ laps}}{26 \text{ min}} = \dfrac{L}{45 \text{ min}}$ $26L = 30 \cdot 45$ $26L = 1350$ $\dfrac{26L}{26} = \dfrac{1350}{26}$ $L = 51.92$

4. a. b. and c. See the image on the right.

5. We can check if the point $(-2, 2)$ fulfills the equation $y = -(1/2)x + 1$:

$2 \overset{?}{=} -(1/2)(-2) + 1$

$2 = 1 + 1$

It does, so the point <u>is</u> on the line.

6. We can write the proportion $x/3.5 = 4/5$, from which $x = (4/5) \cdot 3.5 = 2.8$.
The unknown side is <u>2.8 m long</u>.

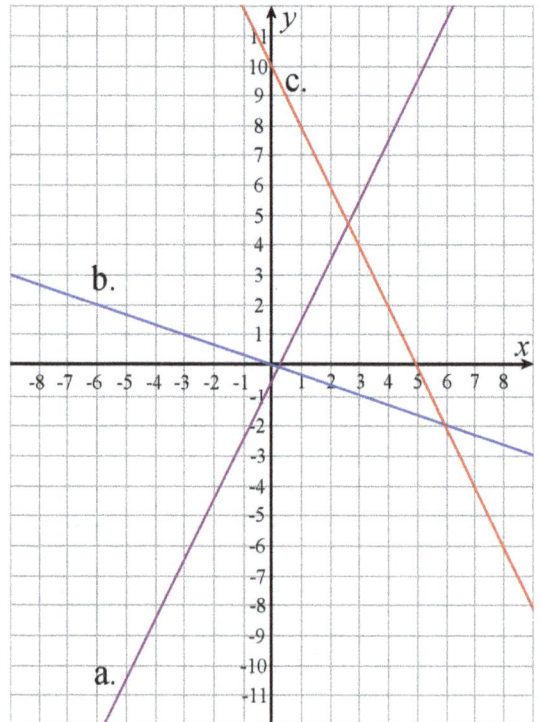

7. a.

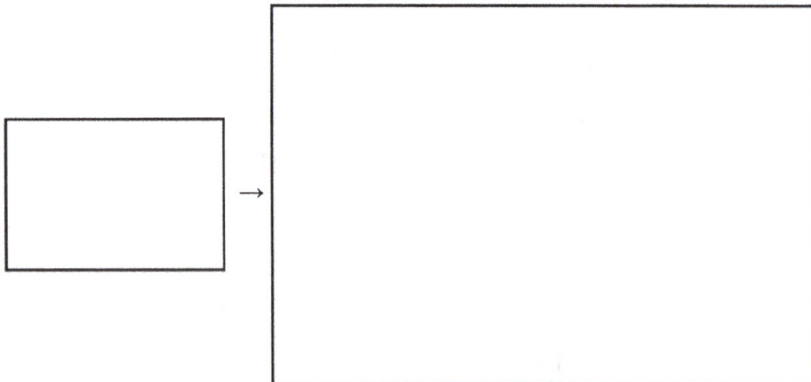

b. The scale ratio is $1{:}2.5 = \underline{2{:}5}$.

198

8. a.

Value of p	$2p$	$p-4$	Value of p	$2p$	$p-4$	Value of p	$2p$	$p-4$
−6	−12	−10	−2	−4	−6	2	4	−2
−5	−10	−9	−1	−2	−5	3	6	−1
−4	−8	−8	0	0	−4	4	8	0
−3	−6	−7	1	2	−3	5	10	1

b. Yes. When $p = -4$, both $2p$ and $p - 4$ equal −8.
c. When p is −3 or greater, $2p$ more than $p - 4$.
d. For $p = 2$.

9.

a. $10 - \dfrac{5}{6} \cdot 2.7$	b. $0.4 \div \left(\dfrac{2}{9} + \dfrac{1}{3} \right)$
$= 10 - \dfrac{13.5}{6}$	$= 0.4 \div \left(\dfrac{2}{9} + \dfrac{3}{9} \right)$
$= 10 - 2.25 = \underline{7.75}$	$= 0.4 \div \dfrac{5}{9}$
	$= \dfrac{4}{10} \div \dfrac{5}{9}$
	$= \dfrac{4}{10} \cdot \dfrac{9}{5} = \dfrac{18}{25}$

10. a. −32
    b. −86
    c. 50

11.

a. $6 \cdot \underline{(-7)} = -42$	b. $-72 \div \underline{(-9)} = 8$	c. $\underline{48} \div (-12) = -4$

12.

a. $1 - x^2$	b. $10xy$	c. $-3(x + y)$
$= 1 - (-5)^2 = 1 - 25 = \underline{-24}$	$= 10(-5)2 = \underline{-100}$	$= -3(-5 + 2) = -3(-3) = \underline{9}$

Geometry Review, p. 80

1. a. Angles v and u, and z and y, are complementary.
   b. Angles w and x are supplementary. So are angles x and y.
   c. Angles w and y are vertical angles.
   d. $u = \underline{31}°$ $x = \underline{129}°$ $y = \underline{51}°$ $z = \underline{39}°$

2. a. $29° + x + 74° = 180°$

   b. $x = 180° - 74° - 29° = \underline{77}°$

3. a. The sum of the measures of the other two angles is $180° - 26° = 154°$. Since the angles are identical, each of the angles measures half of that sum: $154° \div 2 = 77°$.

   b. Check the student's work. The image at the right is not to scale, but it shows the general shape of the triangle. As long as the base measures 4 inches, the sides may be of any length.

4. Answers will vary since the angles chosen may be different. Check the student's work. For example:

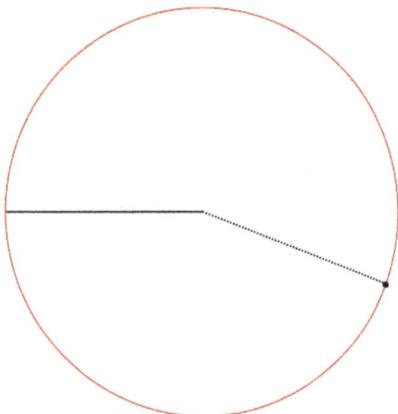

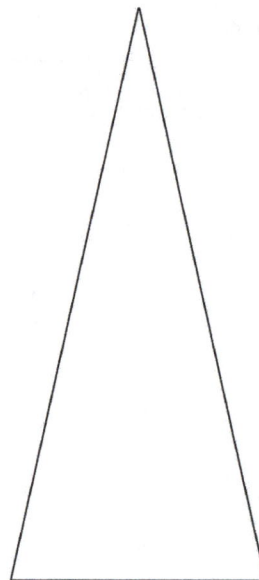

(1) Lightly draw a circle using one of the endpoints of the given line segment as the center and the line segment as the radius.

Choose any point on the circumference of that circle. Draw a line segment from the center to that point as the second side of the triangle.

(2) Complete the figure by drawing in the third side of the triangle. Erase the circle.

5. The image is not to scale.

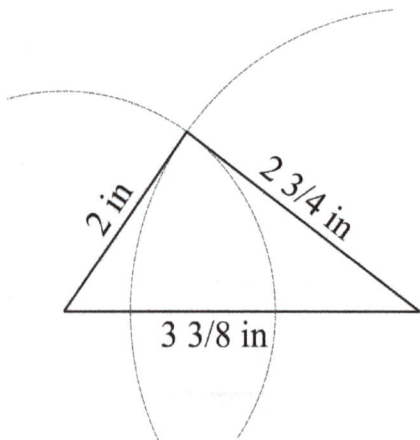

2 in

2 3/4 in

3 3/8 in

6. Yes, the information defines a unique parallelogram. It is true that you can draw it in two different orientations, but those are congruent (you can get one from the other by reflecting and rotating it).

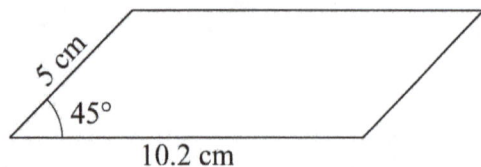

5 cm

45°

10.2 cm

7. The dimensions of the room drawn at a scale of 1:60 are 5.0 cm by 4.2 cm.

8. You can choose any side of the triangle to be the base, and thus you can draw three different altitudes into the triangle. The most natural one might be this:

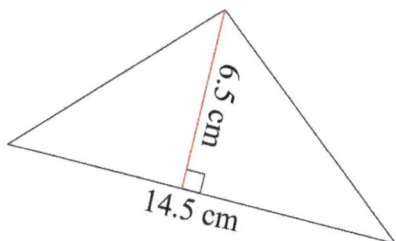

The area of the triangle is then A = 14.5 cm · 6.5 cm ÷ 2 = 47.125 cm^2 ≈ 47 cm^2.

9. a. Diameter = C/π ≈ 6.55718 cm.
   Radius = 6.55718 cm ÷ 2 = 3.27859 cm ≈ <u>3.3 cm</u>.

   b. Diameter = 8 ft 2 in = 98 in. Radius = 98 in ÷ 2 = 49 in. Area = π · (49 in)^2 ≈ 7,540 in^2.

10. The pictures show that if we divide a circle into several sectors, we can rearrange those sectors to form a shape that is very close to a parallelogram. The base of the parallelogram is half of the circumference of the circle, or ½C. The altitude of the parallelogram is the radius of the circle, or r. So the area of the parallelogram is ½C · r, and that is also the area of the circle.

11. The surface area consists of four identical triangles. It is A = 4 · 5.5 ft · 4.6 ft ÷ 2 = 50.6 ft^2 ≈ <u>51 ft^2</u>.

12. The edges of the prism, in inches, are: 2 ft = 24 in, 4 ft = 48 in, and 3 ft = 36 in.
    The volume is then V = 24 in · 48 in · 36 in = 41,472 in^3 ≈ <u>41,500 in^3</u>.

13. a. A = (90 ft + 210 ft)/2 · 150 ft = <u>22,500 ft^2</u>

    b. One square yard = (3 ft) · (3 ft) = 9 ft^2. We can convert the area 22,500 ft^2 into square yards by multiplying it by the conversion factor 1 yd^2/9 ft^2, which essentially means we divide it by 9: A = 22,500 ft^2 · 1 yd^2/9 ft^2 = <u>2,500 yd^2</u>.

14. a. 530 ml = 530 cm^3.

    b. If we use his measurements for the diameter and the height, we get an approximate volume of V = π · (4.5 cm)^2 · 12 cm ≈ 760 cm^3, which is very different from 530 cm^3. So either the diameter or the height is incorrect.

15. a. V = (24 mm + 44 mm)/2 · 32 mm · 230 mm = 250,240 mm^3 ≈ 250,000 mm^3

    b. V = (2.4 cm + 4.4 cm)/2 · 3.2 cm · 23 cm = 250.24 cm^3 ≈ 250 cm^3, or you can take the volume in cubic millimeters and divide it by 10^3 = 1,000 to get the same result.

16. The cross-section is a rectangle.

17. There are two ways:

    (1) Any plane that passes through the top vertex will make a triangular cross section.

    (2) Any plane that passes through just one of the four corners of the pyramid next to the base will also make a triangular cross section.

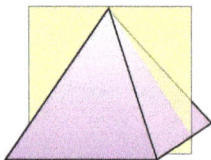

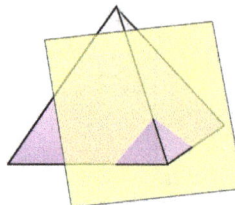

Geometry Test, p. 88

1. $x = 180° - 128° = \underline{52°}$ $y = \underline{128°}$ $z = x = \underline{52°}$

2. Drawings will vary. Check the student's drawing. The measures of the two angles should sum to 90°. The two angles don't have to be adjacent, though most likely the student will draw them so. For example, see the image on the right.

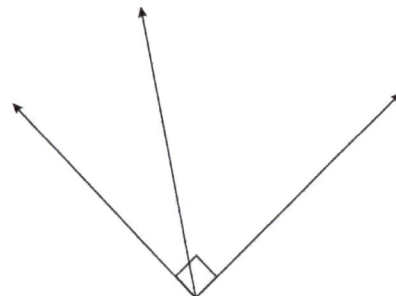

3. Equation for x: $56° + x + 74° = 180°$

   Solution: $x = 180° - 56° - 74° = \underline{50°}$

4. In triangle ABC, the angle at C is 55° (vertical angles). Then, because the angle sum of any triangle is 180°, we get $\beta = 180° - 55° - 72° = \underline{53°}$.

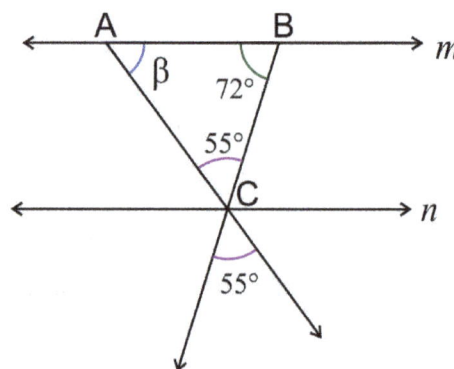

5. a. The base angles measure $(180° - 80°)/2 = \underline{50°}$.

   b. See the image below. The student's triangle may be in a different orientation but should be congruent to this triangle.

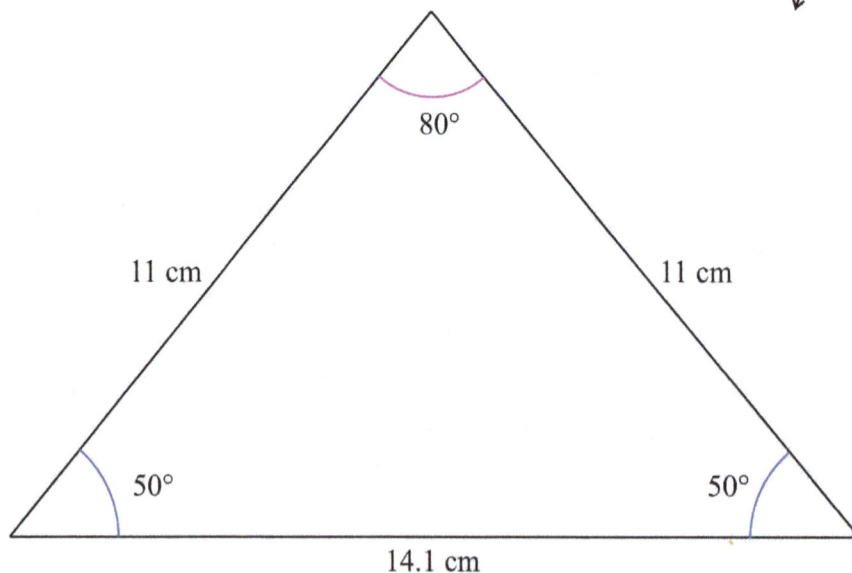

6. The drawings will vary. Check the student's drawing. The two lines should be at 90-degree angles. For example:

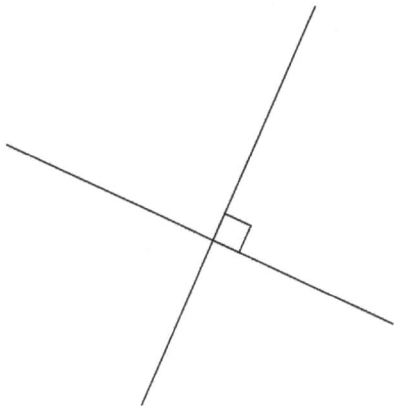

7. The sides of the triangle measure 8.6 cm or 3 3/8 in = 3.375 in. However, if you printed the pages with a setting such as "print to fit" or "shrink to fit," please check the student's work as the student's measurements will differ from those.

   The triangle was drawn at a scale of 1:10, which means that in reality, the sides of the triangle are 86 cm or 33.75 in. When drawn at the scale of 1:12, the sides become 86 cm ÷ 12 ≈ 7.2 cm or 33.75 in ÷ 12 = 2.8125 in = 2 13/16 in.

   Scale 1:10 — the sides are 8.6 cm or 3 3/8 in.   Scale 1:12 — the sides are 7.2 cm or 2 13/16 in.

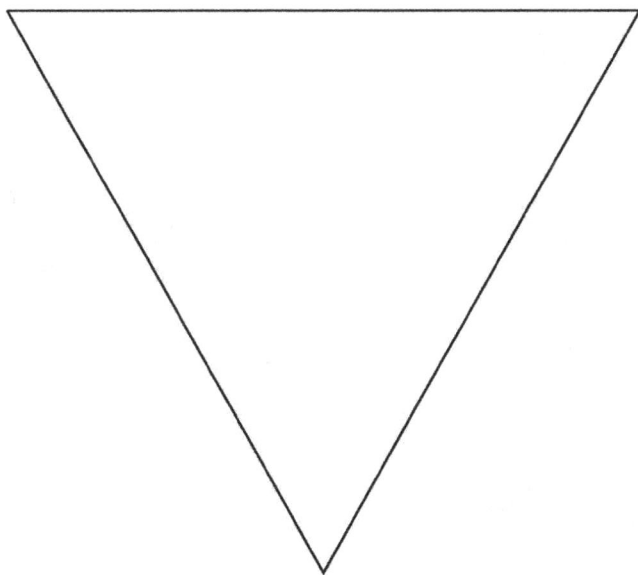

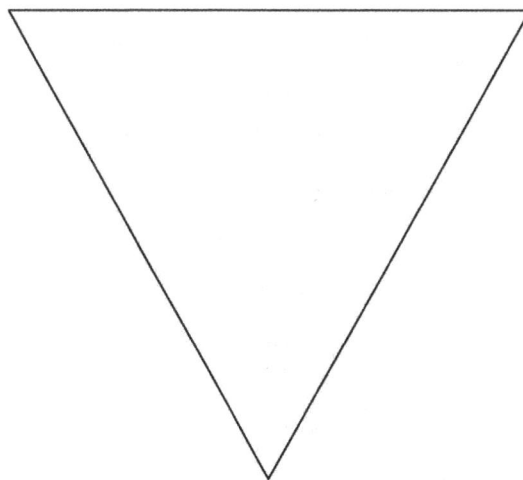

8. a. The area is A = $(a + b)/2 \cdot h$ = (500 ft + 900 ft)/2 · 550 ft = <u>385,000 sq. ft.</u>

   b. A = 385,000 sq. ft. · 1 acre / 43,560 sq. ft ≈ <u>8.84 acres</u>

9. It is a rectangle.

10. It is a rectangle.

11. The area is A = π · (20 cm)^2 ≈ 1,256.6370614 cm^2 ≈ <u>1,260 cm^2</u>.

12. In meters, the edges of the prism measure 0.8 m, 0.4 m, and 0.4 m. Its volume is
    V = 0.8 m · 0.4 m · 0.4 m = <u>0.128 m^3</u>.

13. The area of the top face is π · (12 cm)^2 ≈ 452.3893 cm^2.
    The length of the face that wraps around the cylinder is π · 24 cm ≈ 75.3982 cm.
    The area of the face that wraps around the cylinder is 75.3982 cm · 30 cm = 2,261.946 cm^2.

    The total surface area is 2 · 452.3893 cm^2 + 2,261.946 cm^2 = 3,166.7246 cm^2 ≈ <u>3,170 cm^2</u>.

1. Answers will vary. Check the student's answer.
   For example: Anna owed her mom $30. Then she borrowed $12 more from her. Now, Anna owes $42.
   Or, a submarine was 30 ft below the surface of water. Then it sank 12 ft more. Now it is at 42 ft below the surface.

2. $53 + (-91) + 21 + (-3) + (-55) = 74 + (-149) = \underline{-75}$

3. $\dfrac{9 \text{ baskets}}{12 \text{ shots}} = \dfrac{3 \text{ baskets}}{4 \text{ shots}} = \dfrac{150 \text{ baskets}}{200 \text{ shots}}$

4. You can write the proportion $7/2 = x/16$ or use logical reasoning to get that the unknown side is $7/2 \cdot 16$ cm $= \underline{56 \text{ cm}}$.

5. The new price is $15 - \$1.5 - \$0.75 = \underline{\$12.75.}$

6. Let p be the original price. Then $0.8p = \$100$, from which $p = \$100/0.8 = \underline{\$125}$.
   Another way to solve this is by logical thinking (perhaps accompanied by a bar model). The $100 is 80% or 4/5 of the original price. This means $\$100 \div 4 = \$25$ is 1/5 of the original price. Therefore, the original price is $5 \cdot \$25 = \underline{\$125}$.

7. In 2014, the total revenue was $\$135{,}000 - \$3{,}500 = \$131{,}500$. The revenue that came from the property tax in 2013 was $0.41 \cdot \$135{,}000 = \$55{,}350$. In 2014 this fell to $\$55{,}350 - \$2{,}100 = \$53{,}250$. In 2014, the county got $\$53{,}250/\$131{,}500 = 0.40494 \approx \underline{40.5\%}$ of its revenue from property tax.

8. a. $4w + 13$
   b. $44w^3$
   c. $3c^2d^3$

9. a. Car 1: $d = \underline{20g}$ Car 2: $d = \underline{24g}$
   b. See the image on the right.
   c. Car 1: Slope is 20. Car 2: Slope is 24.
   d. See the graph above. The points are (6, 120) and (5, 120).
   e. See the graph above. The points are (1, 20) and (1, 24).
   f. With zero gallons, the car will travel zero miles.

10. a. $(0.81 - 0.78)/0.78 \approx 0.0384615 \approx \underline{3.8\%}$

    b. $(487 - 445)/445 \approx 0.0943820 \approx \underline{9.4\%}$

11. The sides of the enlarged rectangle measure
    $3.5 \cdot (2\ 1/4$ in$) = 3.5 \cdot 2.25$ in $= 7.875$ in and
    $3.5 \cdot 3$ in $= 10\ 1/2$ in.
    The area is then 7.875 in $\cdot 10.5$ in $= 82.6875$ sq in
    $\approx \underline{82.7 \text{ sq in.}}$

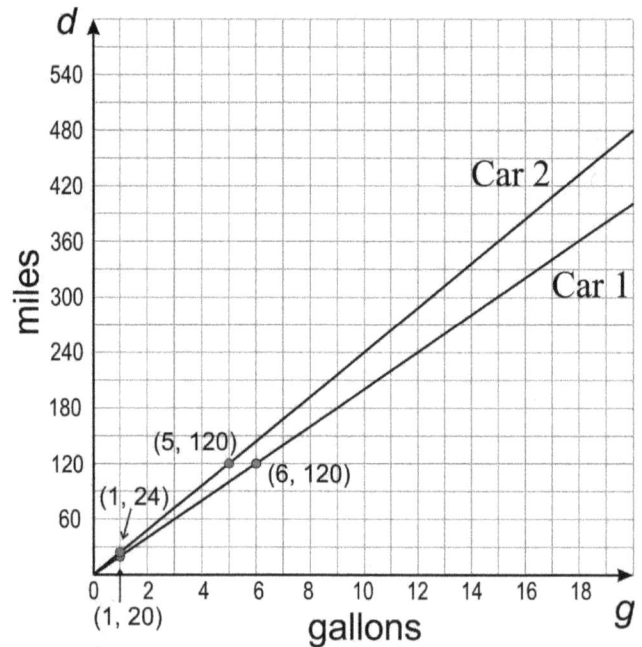

12. $\dfrac{8\ 3/8}{3/4} = 8\dfrac{3}{8} \div \dfrac{3}{4} = \dfrac{67}{8} \cdot \dfrac{4}{3} = \dfrac{67}{2} \cdot \dfrac{1}{3} = \dfrac{67}{6} = 11\dfrac{1}{6}$

    You can cut eleven pieces that are 3/4 ft long out of 8 3/8 feet of string.

13. The sides of the rectangle are 7a and 4b long.

14.

a. $1\dfrac{1}{3} - y = \dfrac{5}{8}$	b. $z + \dfrac{2}{3} = 1\dfrac{9}{10}$
$-y = \dfrac{5}{8} - 1\dfrac{1}{3}$	$z = 1\dfrac{9}{10} - \dfrac{2}{3}$
$-y = -\dfrac{17}{24}$	$z = 1\dfrac{27}{30} - \dfrac{20}{30}$
$y = \dfrac{17}{24}$	$z = 1\dfrac{7}{30}$

Mixed Review 14, p. 95

1. a.

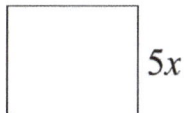

    $5x$ / $6x$

  b. The area is $30x^2$.
  c. The perimeter is $22x$.

2. a. Since the discounted price of $29.97 is 3/5 of the original price, we can get 1/5 of the original price by calculating $29.97 ÷ 3 = $9.99. Then, the original price is 5 · $9.99 = $49.95.

  b. Let p be the original price. Then, $(3/5)p = $29.97.

$$
\begin{aligned}
(3/5)p &= \$29.97 \\
3p &= \$149.85 \\
p &= \$49.95
\end{aligned}
$$

The solution steps for the equation are not exactly the same as the steps for the solution in (a). However, in both solutions we multiply by 5 and divide by 3, just in a different order.

3. a. Discount = ($9 − $8.10)/$9 = $0.90/$9 = 10%.
  b. Percentage of increase = ($26 − $20)/$20 = 6/20 = 30/100 = 30%.

4. His score is $16/21 \approx 76.2\%$.

5. He pays 0.225 · $2,350 = $528.75 in taxes. He has $2,350 − $528.75 = <u>$1,821.25 left</u> after taxes.

6. a. The farmer gets $3/35.2 ≈ $0.08522 for one liter of corn.

  b. P = 0.08522V

  c. Answers will vary because the scaling on the P-axis will vary. Check the student's graph.
    For 100 liters of corn, the farmer earns 100 · $0.08522 = $8.52 and for 600 liters, he earns 600 · $0.08522
    = $51.13. It makes sense to scale the P-axis so that each unit is, say, $8 or $10. In the image below,
    I chose to make each unit on the P-axis to be $10.

6. d. and e.

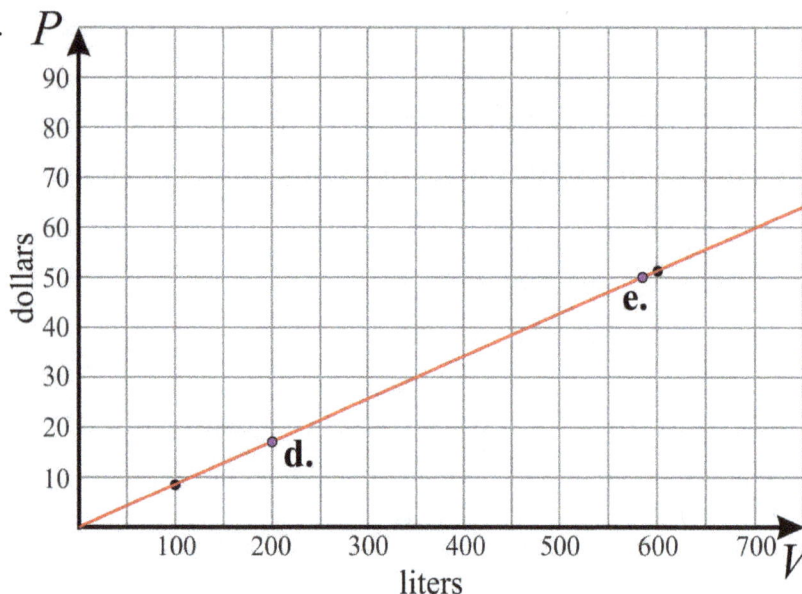

7. a. Yes, her answer is correct.

| A printing press printed 1,500 copies of a book. 5/6 of those were printed as paperbacks and the rest were printed with hard covers. Now, 3/5 of the paperbacks need to be sent to various book stores. *How many books is that?* | Jane's calculation to solve this: $$\frac{\overset{1}{\cancel{3}}}{\underset{1}{\cancel{5}}} \cdot \frac{\overset{1}{\cancel{5}}}{\underset{2}{\cancel{6}}} \cdot 1{,}500 = \frac{1}{2} \cdot 1{,}500 = 750$$ |

b. Yes, the way she calculated it is correct. In the problem, we need to find out 5/6 of 1,500, which can be calculated by multiplying 5/6 and 1,500. Then, we need to find 3/5 of that result, which again can be calculated by multiplying. So in essence, we calculate (3/5) · [(5/6) · 1,500]. Multiplication is associative so we can multiply the first two fractions first, instead of multiplying 5/6 and 1,500 first.

8. a. See the image at the right (not to true scale).
   The other side of the rectangle measures 4.5 cm.

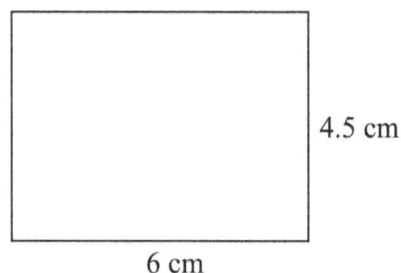

4.5 cm

6 cm

8. b. The sides of the enlarged rectangle are 5/3 = 1 2/3 times the sides of the original rectangle. This means the sides measure (5/3) · 6 cm = 10 cm and (5/3) · 4.5 cm = 7.5 cm. The image on the right is not to true scale.

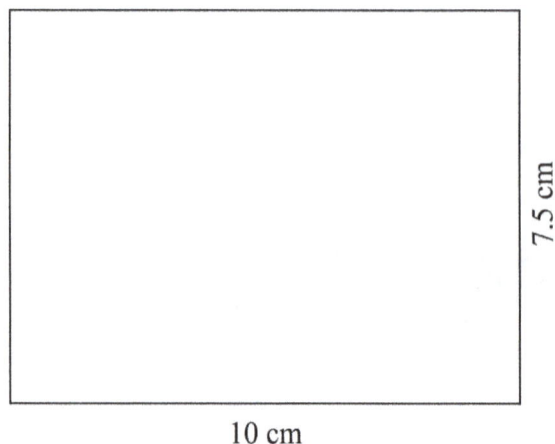

c. It is 4:3. The aspect ratio does not change when a shape is scaled (enlarged or shrunk) with a certain scale ratio.

7.5 cm

10 cm

Pythagorean Theorem Review, p. 98

1. a. 12   b. −9   c. 40   d. 8   e. 49   f. 20

2. a. 7 cm²   b. $\sqrt{20}$ cm

3.

a. $y^2 + 18 = 35$	b. $0.6h^2 = 4$
$y^2 = 17$	$h^2 = 4/0.6 = 40/6 = 20/3$
$y = \sqrt{17} \approx 4.123$	$h = \sqrt{20/3} \approx 2.582$
or $y = -\sqrt{17} \approx -4.123$	or $h = -\sqrt{20/3} \approx -2.582$
Check: $(\sqrt{17})^2 + 18 \overset{?}{=} 35$	Check: $0.6 \cdot (\sqrt{20/3})^2 \overset{?}{=} 4$
$17 + 18 = 35$ ✓	$0.6 \cdot (20/3) \overset{?}{=} 4$
	$(6/10) \cdot (20/3) \overset{?}{=} 4$
	$120/30 = 4$ ✓

4. a. $20^2 + 24^2 \overset{?}{=} 30^2$

$400 + 576 \overset{?}{=} 900$

$976 > 900$

No, they don't form a right triangle. (They would form an acute triangle.)

b. $1^2 + 2.4^2 \overset{?}{=} 2.6^2$

$1 + 5.76 \overset{?}{=} 6.76$

$6.76 = 6.76$

Yes, they form a right triangle.

5. We can ignore the negative answers because a side cannot have a negative length.

a. $s^2 = 3^2 + 5^2$

$s^2 = 9 + 25$

$s^2 = 34$

$s = \sqrt{34} \approx \underline{5.8 \text{ units}}$

b. $x^2 + 21.1^2 = 22.5^2$

$x^2 + 445.21 = 506.25$

$x^2 = 61.04$

$x = \sqrt{61.04} \approx \underline{7.8 \text{ units}}$

6. The pennant is an isosceles triangle. We calculate its altitude using the Pythagorean Theorem. From the right triangle in the image, we get:

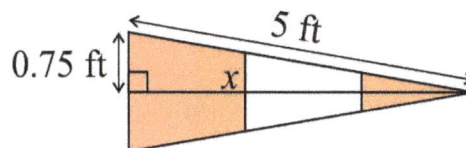

$0.75^2 + x^2 = 5^2$

$0.5625 + x^2 = 25$

$x^2 = 24.4375$

$x = \sqrt{24.4375} \approx 4.94343... \text{ ft}$

So the area is $A = bh/2 \approx 1.5 \text{ ft} \cdot 4.94343 \text{ ft} / 2 = 3.70757 \text{ ft}^2 \approx \underline{3.7 \text{ ft}^2}$.

Pythagorean Theorem Review, cont.

7. Let x be the distance from B to C along 5th Avenue South. Then:

$$370^2 + x^2 = 620^2$$
$$136{,}900 + x^2 = 384{,}400$$
$$x^2 = 247{,}500$$
$$x = \sqrt{247{,}500} \approx 497.49 \text{ m} \approx 500 \text{ m}$$

To go directly from A to C is 620 m, and the distance from A to B and then to C is 370 m + 500 m = 870 m. Therefore, to go directly from A to C is 870 m − 620 m = <u>250 m shorter</u> than to go from A to B and then to C.

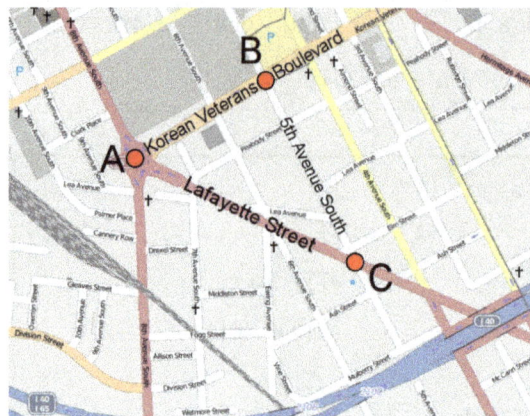

Pythagorean Theorem Test, p. 101

1. a. 1,000   b. 20   c. 1   d. 8   e. 100   f. 53

2. No, they don't: $\sqrt{-9}$ is not a real number, and $-\sqrt{9}$ has the value −3.

3.

a.	b.
$s^2 - 17 = 19$ $s^2 = 36$ $s = \sqrt{36} = 6$ or $s = -\sqrt{36} = -6$	$5y^2 = 89 + 36$ $5y^2 = 125$ $y^2 = 25$ $y = \sqrt{25} = 5$ or $y = -\sqrt{25} = -5$

4. No, they don't:

$$7^2 + 10.2^2 \overset{?}{=} 13.4^2$$
$$49 + 104.04 \overset{?}{=} 179.56$$
$$153.04 < 179.56$$

5. Applying the Pythagorean Theorem to the triangle in the problem, we get:

$$x^2 + 3.32^2 = 3.87^2$$
$$x^2 + 11.0224 = 14.9769$$
$$x^2 = 3.9545$$
$$x = \sqrt{3.9545} \approx 1.99 \text{ m} \quad \text{(We ignore the negative root.)}$$

6. Let x be the length of the diagonal. Applying the Pythagorean Theorem to the triangle formed by the diagonal and the two sides of the square, we get:

$$50^2 + 50^2 = x^2$$
$$2{,}500 + 2{,}500 = x^2$$
$$x^2 = 5{,}000$$
$$x = \sqrt{5{,}000} \approx 70.71 \text{ cm} \quad \text{(We ignore the negative root.)}$$

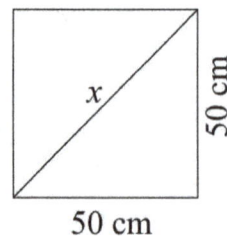

7. The area of the rectangular part is 26 ft · 10 ft = 260 ft^2.

To calculate the area of the top triangle, we have to first find its altitude.
Applying the Pythagorean Theorem to the triangle shown in the image on the right, we get:

$$h^2 + 13^2 = 16.5^2$$
$$h^2 + 169 = 272.25$$
$$h^2 = 103.25$$
$$h = \sqrt{103.25} \approx 10.1612 \text{ ft} \quad \text{(We ignore the negative root.)}$$

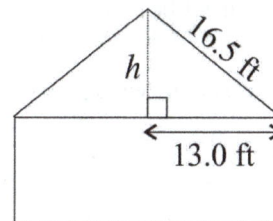

Then, the area of the entire top triangle with 26-ft base is A = 26 ft · 10.1612 ft / 2 = 132.0956 ft^2.

Lastly, the total area is 260 ft^2 + 132.0956 ft^2 = 392.0956 ft^2 ≈ <u>392 ft^2</u>.

Mixed Review 15, p. 104

1. a. Let x be the total area the farmer is planting. Then, $(2/5)x = 3$ acres..

b.
$$(2/5)x = 3 \quad | \cdot 5$$
$$2x = 15 \quad | \div 2$$
$$x = 7.5$$

The farmer is planting 7.5 acres this year.

2. a.

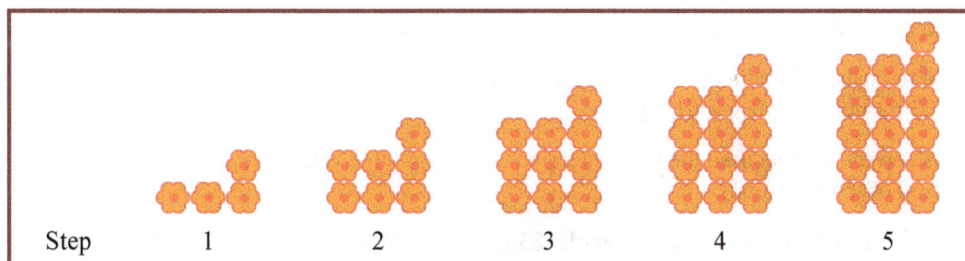

b. The pattern adds a row of three flowers in each step.
c. 39 · 3 + 1 = 118
d. 3n + 1

3. Fill in the missing parts in this justification for the rule *"Negative times negative makes positive."*

(1) Substitute $a = -1$, $b = 1$, and $c = -1$ in the distributive property $a(b + c) = ab + ac$.

$-1(1 + (-1)) = -1 \cdot 1 + (-1) \cdot (-1)$

(2) The whole left side is zero because <u>1</u> + <u>(−1)</u> = 0.

(3) So the right side must equal zero as well.

(4) On the right side, −1 · 1 equals <u>−1</u>. Therefore, −1 · (−1) must equal <u>1</u> so that the sum on the whole right side will equal zero.

4. The percent increase is ($19.95 − $14.95) / $14.95 ≈ 0.3344 ≈ 33.4%.

5. a. The equation is: $x + 117° = 180°$. Solution: $x = 180° − 117° = 63°$.

b. The equation is: $y + 55° = 180°$. Solution: $y = 180° − 55° = 125°$.

6. A = <u>10</u> sq. yd. = <u>90</u> sq. ft.

7. The volume of one glass is $V = A_b h = \pi \cdot (3 \text{ cm})^2 \cdot 8 \text{ cm} \approx 226.195 \text{ cm}^3$.

When filled 3/4 full, the glass contains $(3/4) \cdot 226.195 \text{ cm}^3 \approx 169.646 \text{ cm}^3 = 169.646 \text{ ml}$ of liquid.

Since $1{,}000 \text{ ml} \div 169.646 \text{ ml} \approx 5.89$, from the 1-liter pitcher Jane can fill <u>5 glasses</u> (and most of a 6th).

8. a. The student's parallelogram may be in a different orientation, but it should be congruent to the parallelogram below.

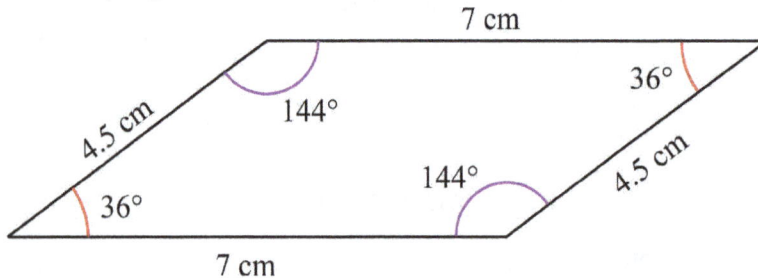

b. To find the area, one needs to draw an altitude to the parallelogram and measure it:

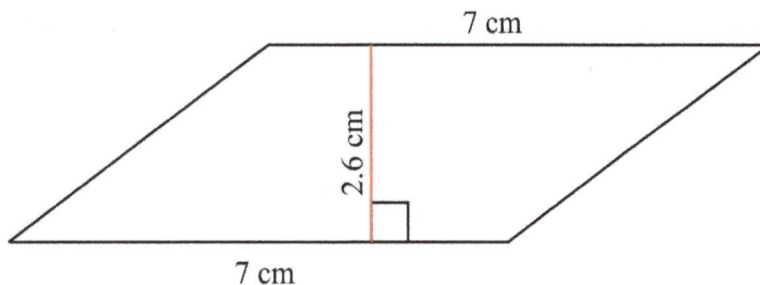

The area is $A = 7 \text{ cm} \cdot 2.6 \text{ cm} = 18.2 \text{ cm}^2 \approx \underline{18 \text{ cm}^2}$.

9. First we convert 200 liters into bushels: 200 liters = 200 L/(35.2 bushels/L) $\approx$ 5.682 bushels. This will earn the farmer 5.682 bushels · $3/bushel = $17.046 ≈ <u>$17.05</u>.

10. a. Mary got both questions correct. Angela used the wrong reference values — the values relating to the *first* package of granola instead of to the second.

b. You can calculate the unit rates for each package of granola: $3.60/0.8 kg = $4.50/kg and $3.00/0.6 kg = $5.00/kg. Or, notice that the first granola package is 33% heavier than the second, yet it is only 20% costlier than the second. This means <u>the first package is cheaper by weight</u>.

11.

a. $8\dfrac{3}{4} \div 2\dfrac{5}{8} + 2\dfrac{2}{5}$

$= \dfrac{35}{4} \cdot \dfrac{8}{21} + 2\dfrac{2}{5}$

$= \dfrac{5}{1} \cdot \dfrac{2}{3} + 2\dfrac{2}{5}$

$= \quad 3\dfrac{1}{3} \quad + 2\dfrac{2}{5}$

$= \quad 3\dfrac{5}{15} \quad + 2\dfrac{6}{15} = 5\dfrac{11}{15}$

b. $4\dfrac{2}{7} \div \dfrac{6}{7} \cdot \dfrac{5}{8}$

$= \dfrac{30}{7} \cdot \dfrac{7}{6} \cdot \dfrac{5}{8}$

$= \dfrac{5}{1} \cdot \dfrac{1}{1} \cdot \dfrac{5}{8} = \dfrac{25}{8} = 3\dfrac{1}{8}$

1. a. Let s be Jayden's salary. The equation is $s/7 = \$415$ or $(1/7)s = \$415$.

b. The equation is $s/7 = \$415$, from which $s = 7 \cdot \$415 = \underline{\$2,905}$.

2.

a. $7x + 21 = 7(x + 3)$	b. $24w - 16 = 8(3w - 2)$
c. $-21t - 7 = -7(3t + 1)$	d. $50a - 70b - 120 = 10(5a - 7b - 12)$
e. $-55a + 30 = -5(11a - 6)$	f. $-56y - 84 - 7x = -7(8y + 12 + x) = -7(x + 8y + 12)$

3. a. $(3\,\tfrac{1}{2}$ cans$)/(2/3$ room$) = 3\,\tfrac{1}{2} \cdot (3/2)$ cans/room $= (7/2) \cdot (3/2)$ cans/room $= 21/4$ cans/room $= 5\,1/4$ cans/room.

b. Ava needs $5\,1/4$ cans to paint the room and she's already used $3\,1/2$ cans of paint.
She needs $5\,1/4 - 3\,1/2 = \underline{1\,3/4 \text{ cans of paint}}$ to finish it.

4. a. $100 - (-2)^2 = 100 - 4 = \underline{96}$

b. $\dfrac{2(1/2)}{1/2 + 3} = \dfrac{1}{3\,1/2} = \dfrac{1}{7/2} = \dfrac{2}{7}$

5.

a. $x - \dfrac{5}{6} = 7\dfrac{1}{3}$ $\quad$ $x = 7\dfrac{1}{3} + \dfrac{5}{6}$ $\quad$ $x = 8\dfrac{1}{6}$	b. $2\dfrac{1}{4} - w = 1\dfrac{2}{7}$ $\quad$ $2\dfrac{1}{4} = w + 1\dfrac{2}{7}$ $\quad$ $2\dfrac{1}{4} - 1\dfrac{2}{7} = w$ $\quad$ $2\dfrac{7}{28} - 1\dfrac{8}{28} = w$ $\quad$ $w = \dfrac{27}{28}$
c. $5y = -\dfrac{4}{9}$ $\quad$ $y = -\dfrac{4}{9} \div 5$ $\quad$ $y = -\dfrac{4}{45}$	d. $v + \dfrac{1}{5} = -\dfrac{1}{12}$ $\quad$ $v = -\dfrac{1}{12} - \dfrac{1}{5}$ $\quad$ $v = -\dfrac{5}{60} - \dfrac{12}{60} = -\dfrac{17}{60}$

6. The percent of increase is $(\$22.50 - \$19)/\$19 \approx \underline{18.4\%}$.

7. The interest was $I = prt = \$1,500 \cdot 0.098 \cdot 1.5 = \$220.50$.
The total amount he paid back was $\$1,500 + \$220.50 = \underline{\$1,720.50}$.

8. a. No, they are not. For example, when the time doubles from 1 hour to 2 hours, the speed does not double, but is halved.

b. There is no need to do anything here, since the quantities were not in proportion.

9. a. Hailey is doing $(602 - 458)/458 \approx 31.4\%$ better than Grace.

b. Tony is doing $(553 - 365)/365 \approx 51.5\%$ better than Chris.

Mixed Review 16, cont.

10. a. Answers will vary. Check that the student's triangle is indeed equilateral.
    b. Answers will vary. Check the altitude the student drew. For an example, see the image on the right.
    c. Answers will vary. Check the student's measurements and area calculation. For example: The area is 4.7 cm · 4.1 cm / 2 ≈ 10 cm².

11. Yes, it does define a unique triangle:

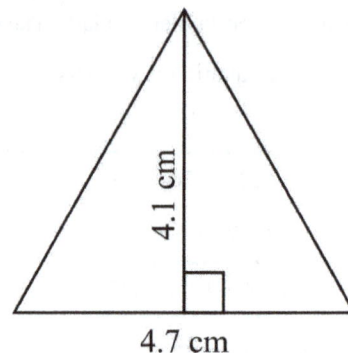

60°   45°

5.6 cm

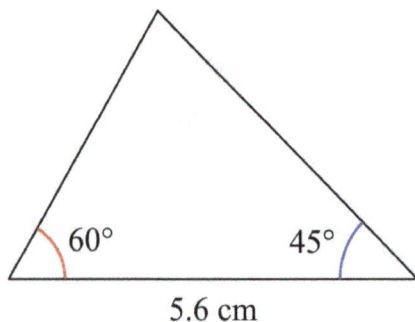

4.1 cm

4.7 cm

12. The volume is ½ · 4.5 cm · 3.9 cm · 24 cm = 210.6 cm³ ≈ 210 cm³.

Probability Review, p. 110

1. a. P(not math, science, or English) = P(social studies, art, or music) = 9/25.
   b. P(math) = 7/25.
   c. There are 13 boys, and four of them chose math. So P(a boy's favorite subject is math) = 4/13.
   d. There are 12 girls, and three of them chose math. So P(a girl's favorite subject is math) = 3/12 = 1/4.

2. You can find these probabilities by listing and counting the favorable outcomes, or since these events are compound events, you can also find them by multiplying the probabilities of the individual events.

   a. P(5, 6) = (1/6) · (1/6) = 1/36.

   b. There are 9 favorable outcomes: (2, 2), (2, 4), (2, 6), (4, 2), (4, 4), (4, 6), (6, 2), (6, 4), and (6, 6), so P(even, even) = 9/36 = 1/4.

      Or: P(even, even) = (1/2) · (1/2) = 1/4.

   c. There are 4 favorable outcomes: (5, 5), (5, 6), (6, 5), and (6, 6), so P(at least 5, at least 5) = 4/36 = 1/9.

      Or: P(at least 5, at least 5) = (2/6) · (2/6) = 4/36 = 1/9.

   d. There are 6 favorable outcomes: (1, 1), (1, 2), (2, 1), (2, 2), (3, 1), and (3, 2), so P(at most 3, at most 2) = 6/36 = 1/6.

      Or: P(at most 3, at most 2) = (3/6) · (2/6) = (1/2) · (1/3) = 1/6.

3. a. P(1) = 8/60 = 2/15. P(4) = 13/60.

   b. Because rolling a die is a chance or random process: you never know what you will get when you roll it. Rolling a die six times does not guarantee that you get one of each number. As the number of repetitions increases, the relative frequencies (experimental probabilities) do get closer and closer to the theoretical probabilities of 1/6. However, 60 is not a large number of repetitions, so we expect the experimental probabilities to vary a lot from 1/6.

Probability Review, cont.

4. a.

Outcome	Frequency	Experimental Probability (%)	Theoretical Probability (%)
HH	38	19%	25%
HT	53	26.5%	25%
TH	46	23%	25%
TT	63	31.5%	25%
TOTALS	**200**	**100%**	**100%**

b. The experimental probabilities would be much closer to the theoretical ones (much closer to 25%).

5. You can find these probabilities by listing and counting the favorable outcomes or by multiplying the probabilities of the individual events. There are a total of $7 \cdot 6 = 42$ possible outcomes, each being equally likely.

a. There is only one favorable outcome (purple, orange), so the probability is 1/42.
Or: P(purple, orange) = $(1/7) \cdot (1/6) = 1/42$.

b. There are five favorable outcomes: (red, blue), (red, purple), (red, orange), (red, yellow), and (red, mint), so the probability is 5/42.
Or: P(red, not pink) = $(1/7) \cdot (5/6) = 5/42$.

c. There are 30 favorable outcomes: each of the six colors *red, blue, purple, pink, orange,* and *yellow* combined with five of those colors so that the same color is not chosen twice. Thus the probability is $30/42 = 5/7$.
Or: P(not mint, not mint) = $(6/7) \cdot (5/6) = 5/7$.

Probability Test, p. 112

1. a. P(not 5) = 5/6

b. P(2 or 6) = 2/6 = 1/3

c. P(less than 9) = 1

d. P(not 2 nor 5) = 4/6 = 2/3

2. a. See the tree diagram on the right. Listing the outcomes at the bottom is totally optional.

b. P(yellow; purple) = 1/12

c. P(red or yellow; orange) = 2/12 = 1/6

d. P(not red; not orange) = 6/12 = 1/2

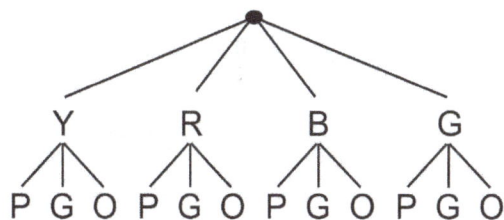

Outcomes: YP YG YO RP RG RO BP BG BO GP GG GO

3. There are 36 possible outcomes, as shown in the grid on the right:

a. The favorable outcomes are: (1, 5), (2, 4), (3, 3), (4, 2), and (5, 1).
The probability of getting a sum of six is 5/36.

b. The favorable outcomes are: (1, 1), (1, 2), (2, 1), and (2, 2).
The probability is 4/36 = 1/9.

c. The favorable outcomes are: (6, 1), (6, 2), (6, 3), (6, 4), (6, 5), and (1, 6), (2, 6), (3, 6), (4, 6), and (5, 6). The probability is 10/36 = 5/18.

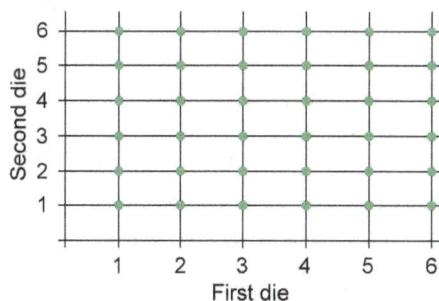

4. a. heads & tails, tails & tails, tails & heads, and heads & heads.
   In abbreviated form: HT, TT, TH, and HH.

b.

	Frequency	Experimental probability	Theoretical probability
TT	5	1.3%	25%
TH	8	2%	25%
HT	182	45.5%	25%
HH	205	51.3%	25%
TOTALS	400	100.6%	100%

c. Perhaps the first coin was weighted so that it almost always lands on heads.

5. It is 1/8, because there are eight equally likely outcomes: HHH, HHT, HTT, HTH, THH, THT, TTT, and TTH.

6. a. P(cat) = 11/120 ≈ 9.2%.
   b. They can expect to get the bear about (17/120) · 300 ≈ 43 times.

Mixed Review 17, p. 115

1. Beth set up the proportion incorrectly. She ends up cross-multiplying cost by cost and liters by liters, which gives her impossible units of $L^2/\$$. Some correct ways to write the proportion are 80/35 = 52/C or \$80/52 = \$35/C. There are other ways as well. The crucial point is that after cross-multiplying, you should get 80C on one side of the equation and 35 · 52 on the other.

Eighty liters of blueberries costs \$35. How much would 52 liters cost?	Eighty liters of blueberries costs \$35. How much would 52 liters cost?
Beth's Answer: 52 liters would cost \$118.86.	Correct Answer: 52 liters would cost \$22.75.
Beth's Solution: $\dfrac{80}{35} = \dfrac{C}{52}$	Correct Solution: $\dfrac{80}{35} = \dfrac{52}{C}$
$35C = 80 \cdot 52$	$80C = 35 \cdot 52$
$35C = 4160$	$80C = 1{,}820$
$\dfrac{35C}{35} = \dfrac{4160}{35}$	$\dfrac{80C}{80} = \dfrac{1{,}820}{80}$
$C = 118.86$	$C = 22.75$

2. $52° + 90° + x = 180°$

   $x = 180° - 52° - 90°$

   $x = 38°$

3. $11.4^2 + 15.2^2 \overset{?}{=} 19^2$

   $129.96 + 231.04 \overset{?}{=} 361$

   $361 = 361$

   Yes. Because the sides satisfy the Pythagorean Theorem, they form a right triangle.

4. $V = 50 \text{ cm} \cdot 50 \text{ cm} \cdot 50 \text{ cm} = 125{,}000 \text{ cm}^3$

   $V = 0.5 \text{ m} \cdot 0.5 \text{ m} \cdot 0.5 \text{ m} = 0.125 \text{ m}^3$

5. a. It is an equilateral triangle.

   b. First, we need to calculate the altitude of the triangle. Since the altitude splits the triangle into two right triangles, we can use the Pythagorean Theorem to calculate the altitude.

   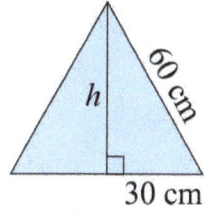

$$h^2 + 30^2 = 60^2$$
$$h^2 + 900 = 3600$$
$$h^2 = 2700$$
$$h = \sqrt{2700}$$
$$h \approx 51.9615 \text{ cm}$$

   So the area is $A = bh/2 = 60 \text{ cm} \cdot 51.9615 \text{ cm} / 2 \approx \underline{1{,}560 \text{ cm}^2}$.

6. a. $A = \pi r^2 = \pi \cdot (7.5 \text{ cm})^2 = 56.25\pi \text{ cm}^2 \approx 177 \text{ cm}^2$.

   b. It occupies $176.7146 \text{ cm}^2 / (21 \text{ cm} \cdot 29.7 \text{ cm}) \approx \underline{28.3\%}$.

7. a. The ratio 2:5 means that the sides of the enlarged rectangle are $5/2 = 2.5$ times the sides of the original rectangle. So the sides become $2.5 \cdot 3\frac{1}{2} \text{ in} = \underline{8\frac{3}{4} \text{ in}}$ and $2.5 \cdot 2 \text{ in} = \underline{5 \text{ in}}$.

   b. The area is $8\frac{3}{4} \text{ in} \cdot 5 \text{ in} = 35/4 \text{ in} \cdot 5 \text{ in} = 175/4 \text{ in}^2 = \underline{43\frac{3}{4} \text{ in}^2}$.

8. a. $7(x + 8) = 7x + 56$

   b. $4(2y - 10) = 8y - 40$

   c. $0.1(2x + 18) = 0.2x + 1.8$

9.

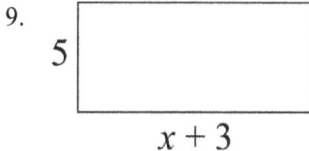

   Other possibilities include a rectangle with sides 1 and $5x + 15$, with sides 2 and $2.5x + 7.5$, with sides 3 and $(5/3)x + 5$, and others that use fractional or decimal lengths of sides.

10.

a.	b.
$\dfrac{v-6}{7} = -31$	$\dfrac{x}{4} - 1 = -5$
$v - 6 = -217$	$\dfrac{x}{4} = -4$
$v = -211$	$x = -16$

11. The change from May to June is measured in terms of May's sales: (*difference*)/(*May's sales*). So the book sales decreased by $(2{,}400 - 2{,}000)/2{,}400 = 400/2400 = 1/6 = 16.7\%$.

12. a. $(77 - 66)/66 = 16.7\%$.

    b. $(86 - 66)/66 = 30.3\%$.

    c. $(86 - 77)/77 = 11.7\%$.

13. a. 9   b. −3   c. −80   d. 6   e. −1/6   f. −1

215

1. a. 1   b. 8   c. 100   d. 20

2. a. elevation = $11n - 9$

b. When $n = 47$, the elevation is $11 \cdot 47 - 9 = 508$ ft.

c.
$$
\begin{aligned}
11n - 9 &= 288 \\
11n &= 297 \\
n &= 27
\end{aligned}
$$

The 27th floor is at a height of 288 ft.

3. First we cut the circle into equal sectors, in this case into 12. Then we rearrange the sectors to form a shape that is close to a parallelogram.

The height of this parallelogram is the radius of the circle (r), and the base of the parallelogram is half of the circumference of the circle ($\frac{1}{2}C$).

The area of the parallelogram is A = $\frac{1}{2}C \cdot r$.

By dividing the circle into a larger number of sectors, the shape made of the sectors would get closer to a parallelogram. In fact, we can get as close to a parallelogram as we want by dividing the circle into a very large number of sectors. So the area of the circle is also A = $\frac{1}{2}C \cdot r$.

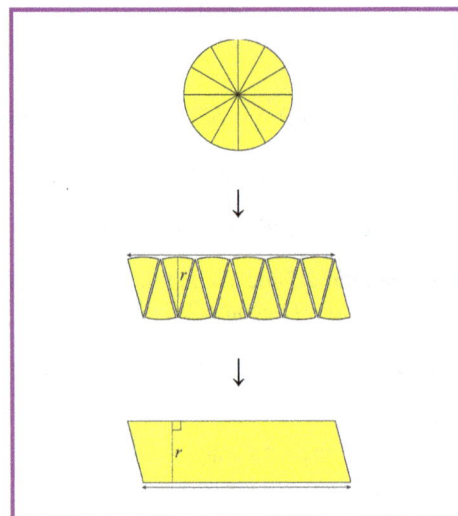

4. a. 10 problems/3 coins = 3 1/3 problems per coin.

b. There are many ways to solve this problem. One is to write a proportion:

$$
\begin{aligned}
\frac{10 \text{ problems}}{3 \text{ coins}} &= \frac{x \text{ problems}}{75 \text{ coins}} \\
3x &= 10 \cdot 75 \\
3x &= 750 \\
\frac{3x}{3} &= \frac{750}{3} \\
x &= 250
\end{aligned}
$$

You need to solve <u>250 problems</u> in order to earn 75 coins.

c. Notice that the game only awards you coins once you complete each *set of 10 problems*, so actually we need to do this calculation using the quantity 370 problems instead of 376 problems. There are many ways to find out the answer. One way is to multiply the given amount, 370 problems, by the unit rate 3 coins/10 problems: 370 problems · (3 coins/10 problems) = <u>111 coins</u>.

If you use the quantity 376 problems, you will get 376 problems · (3 coins/10 problems) = 112.8 coins, but this is a wrong answer since the game only gives the coins in sets of 3 after a set of 10 problems has been completed.

5. The circular clock has an area of A = $\pi \cdot (18 \text{ cm})^2 \approx 1{,}018 \text{ cm}^2$.
The square clock as an area of $(30 \text{ cm})^2 = 900 \text{ cm}^2$.

The <u>circular clock</u> covers $1{,}018 \text{ cm}^2 - 900 \text{ cm}^2 = \underline{118 \text{ cm}^2}$ more of the wall.

6. Using 75% of her quota in 18 days means that she used 150 gigabytes in 18 days, and that gives us a daily rate of 150 Gb/18 days = 8 1/3 Gb/day.

Since she used 75% of her quota in 18 days, she will use the remaining 25% in six more days, totaling 24 days to use the 200Gb. Then in the last six days of September she will use 6 days · 8 1/3 Gb/day = 50 Gb, and that will cost her 50 Gb · $0.35 = <u>$17.50 extra to pay</u> at the end of the month.

7. We can write a proportion $84/108 = 70/x$ or the proportion $70/84 = x/108$ or even just use logical reasoning: Since the shorter side is $70/84 = 5/6$ of the longer side of the pentagon, then in the larger pentagon, the unknown side is $(5/6) · 108$ mm = <u>90 mm</u>.

8.
$$24^2 + 11.1^2 = x^2$$
$$576 + 123.21 = x^2$$
$$x^2 = 699.21$$
$$x = \sqrt{699.21} \approx \underline{26.4 \text{ ft}}$$

9. a. The area of the top and bottom bases is $2 · 16.2 \text{ cm}^2 = 32.4 \text{ cm}^2$. Then, the area of each rectangular face is $12 \text{ cm} · 2.5 \text{ cm} = 30 \text{ cm}^2$. The total surface area is $32.4 \text{ cm}^2 + 6 · 30 \text{ cm}^2 = 212.4 \text{ cm}^2$.

   b. The volume is $V = A_b h = 16.2 \text{ cm}^2 · 12 \text{ cm} = \underline{194.4 \text{ cm}^3}$.

10. The possible numbers to make are: 22, 23, 25, 27, 32, 33, 35, 37, 52, 53, 55, 57, 72, 73, 75, and 77 - a total of 16 possible outcomes.

   a. P(between 31 and 40) = 4/16 = 1/4

   b. P(both digits same) = 4/16 = 1/4

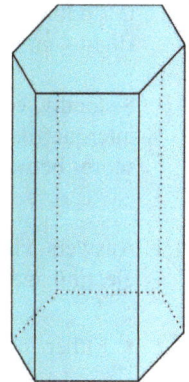

Statistics Review, p. 121

1. a. Here is the data for the number of baskets Jake shot in order: 26 27 27 29 30 31 33 33 34 35

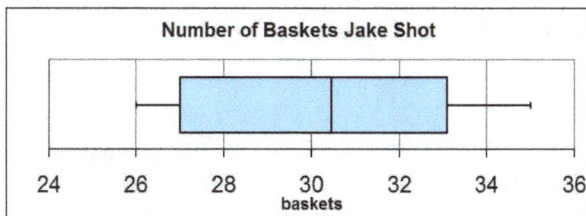

Number of Baskets Jake Shot

   b. The median and mean both are 30.5. That would mean that Jake's percentage of successful baskets is 30.5/50 = 61%. Using that percentage, we can estimate that Jake will make 0.61 · 120 ≈ 73 baskets in 120 shots.

2. a. Harry's sampling method is not random. The people who stay after the meeting are probably not representative of the club membership as a whole.

   b. Answers will vary. Check the student's answer. For example: Harry could get an alphabetical list of the club members. He could choose a random number between 1 and 6 by rolling a die. Then he would choose the member with that number from the list plus every member after that at some regular interval, such as every 4th or 6th member, depending on how many people he needs for his survey.

3. a. See the stem-and-leaf plot on the right.

   b. Grade 7 appears to have done better in the test.
      We cannot tell from the plot which Group has a greater variability.

   c. <u>Grade 7:</u>   Median <u>75.5</u>   Range <u>41</u>

      Interquartile range: <u>83.5 - 67 = 16.5</u>

      <u>Grade 8:</u>   Median <u>68.5</u>   Range <u>47</u>

      Interquartile range: <u>75.5 - 61 = 14.5</u>

<table>
<tr><th colspan="3">Test results</th></tr>
<tr><th>SAMPLE 1
GRADE 7
Leaf</th><th>Stem</th><th>SAMPLE 2
GRADE 8
Leaf</th></tr>
<tr><td>9 6</td><td>5</td><td>2 4 5 8</td></tr>
<tr><td>8 8 6 4 1</td><td>6</td><td>0 2 2 3 5 8 9</td></tr>
<tr><td>9 9 7 6 5 5 3</td><td>7</td><td>0 2 2 4 7</td></tr>
<tr><td>8 4 3</td><td>8</td><td>2 5</td></tr>
<tr><td>7 6 0</td><td>9</td><td>2 9</td></tr>
</table>

7 | 8 means 78

   d. Yes, they do. First of all, the median for grade 7 is more than the median for grade 8. The difference in medians is 7 points. This is about half of the interquartile range of either group, so the difference is not significant, only slight—and that is what we could see in the graph: that 7th graders seem to have done better, but not by much.

      Secondly, concerning variability: Grade 8 has a somewhat larger range, but not by much. Grade 7 has a larger interquartile range but not by much. Based on all that, there is not a huge difference in the variability in the test scores between the two groups.

4. a. Answers will vary. Check the student's answers. The student should include a statement about the winner of the election and at least 2 other inferences. For example:

   • Either Johnson or Garcia will win, but we cannot tell which one from these two survey results.

   • Most likely, Wilson will come in third place, Evans fourth place, and Hanley fifth place.

   • Wilson will probably get a little more than half the votes that Johnson or Garcia each get.

   • Evans will get less than half the votes that Johnson or Garcia will get.

   • Hanley will get only a small fraction of the votes that Johnson or Garcia will get.

   • Both Johnson and Garcia will get almost 1/3 of the total vote.

   b. The average percentage of people who voted for Wilson is $((13 + 16) \div 2) / 75 = 14.5/75 = 19.\overline{3}\%$. Using that, we estimate that $0.19\overline{3} \cdot 1{,}230 \approx 240$ students will vote for Wilson.

      If we use the result from Survey 1 where Wilson got 13/75 of the total vote, we'd estimate that he would get $(13/75) \cdot 1{,}230 \approx 213$ votes. And if we use the result from Survey 2, we'd estimate that he would get $(16/75) \cdot 1{,}230 \approx 262$ votes. These two quantities (213 and 262) differ from our actual estimate of 240 by a few dozen (or more than 20) votes. Since the samples are random, the estimate of 240 votes may be off by 2-3 times that amount; we cannot really know without having many more samples.

      So, the estimate of 240 votes may be <u>off by several dozens of votes.</u> You could also say that it may be off by 50 or so votes.

5. a. Store 1 appears to have cheaper prices.
      Neither store appears to have greater variability in the prices. Both the ranges and the interquartile ranges are fairly similar.

   b. The difference in the prices is not significant. The difference in the medians is about $20, and the interquartile ranges are about $100. Since the difference in the medians is only about 1/5 of the interquartile range, it is not significant.

1. a. Group 1 appears to have lost more weight.
   b. We cannot tell from the plot. The ranges and the interquartile ranges seem quite similar for both groups.
   c. It means the person actually *gained* 1 pound.
   d. Yes, method 1 is better. The difference in the medians is 11 lb - 6 lb = 5 lb. The interquartile ranges are about 4 lb. This means the difference in the medians is more than 1 time the interquartile range, which makes it a significant difference.

2.

Sampling method	Biased or not?
(1) Erica generates 120 random numbers between 1 and 1,000, and chooses the corresponding people that she meets on the main street of the town. If method (1) is biased, explain why: Most likely, not every person in this town tends to visit the main street on foot. So, not everyone in this town has an equal chance of being selected in Erica's sample.	Yes.
(2) Erica chooses randomly 120 people from a list of the town's residents. If method (2) is biased, explain why:	Not.
(3) Erica places a box and her survey papers in the town's library, and anyone who wants to can fill in the survey questionnaire. If method (3) is biased, explain why: There are people who never visit the library so they don't have a chance to be selected in the sample	Yes.
(4) Erica chooses randomly 120 people from a list of the town's taxpayers. If method (4) is biased, explain why: Some people may not have to pay any taxes. Those people would not be in the list of taxpayers. Those people don't have a chance to be selected in the sample.	Yes.

3. Answers will vary. Check the student's answer. The student should produce at least 4 inferences to get the full 4 points. For example, we can infer these things:

   - Cappuccino is the most wanted new flavor.
   - Peanut Butter, Pineapple, and Blueberry come in the second place, and they are all approximately equally popular as new flavors.
   - Kiwi is the least popular choice for a new flavor.
   - Cappuccino got more than twice the "votes" that Pineapple, Peanut Butter, or Blueberry got.
   - Cappuccino was the favorite of nearly half of the people.
   - About 1/6 of the customers would like either Pineapple, Peanut Butter, or Blueberry added to the selection.
   - Only a small fraction of the customers would care for Kiwi to be added.

4. a. The means differ by 28.6 years − 27.55 years = 1.05 years.

   b. No, the difference is not significant, because the difference of 1.05 years is only a small fraction of the mean absolute deviation (6.78 or 6.205 years).

1. The correct equation is $\dfrac{4p}{5} = \$9.40$.

2. The original price of the rubber boots was $11.75.

   At the right you will find the most common way to solve the equation (where we first multiply both sides by 5, then divide both sides by 4). Another way to solve it would be to divide both sides by the fraction 4/5.

$$\dfrac{4p}{5} = \$9.40 \qquad \Big| \cdot 5$$

$$4p = \$47 \qquad \Big| \div 4$$

$$p = \$11.75$$

3. We calculate the final price of the juicer in two steps. After the first discount, the juicer costs $0.85 \cdot \$200 = \$170$. After the second discount, it costs $0.8 \cdot \$170 = \$136$. If it had been discounted from $200 to $136 in one step, the percentage of discount would have been $(\$200 - \$136)/\$200 = \$64/\$200 = 32/100 = \underline{32\%}$.

4. a. The volume of the first can is $V = A_b h = \pi \cdot (3.3 \text{ cm})^2 \cdot 8.5 \text{ cm} \approx 290.8015 \text{ cm}^3 \approx \underline{291 \text{ cm}^3}$.
   The volume of the second can is $V = A_b h = \pi \cdot (5 \text{ cm})^2 \cdot 5.8 \text{ cm} \approx 455.5309 \text{ cm}^3 \approx \underline{455 \text{ cm}^3}$.

   b. The larger can is $(455.5309 \text{ cm}^3 - 290.8015 \text{ cm}^3)/290.8015 \text{ cm}^3 \approx \underline{56.6\%}$ bigger than the smaller can.

5. To get the true dimensions, we multiply each of the dimensions in the plan by the ratio 6 ft/1 in:

   5 ¼ in · 6 ft/1 in = $\underline{31 \ 1/2 \text{ ft}}$ and   6 ¾ in · 6 ft/1 in = $\underline{40 \ 1/2 \text{ ft}}$.

6. a. and b. See the image on the right.

   c. If angle A is 102°, then supplementary angle C is 180° − 102° = 78°.

   d. The quadrilateral enclosed by two pairs of equal-length parallel lines is called a <u>parallelogram</u>.

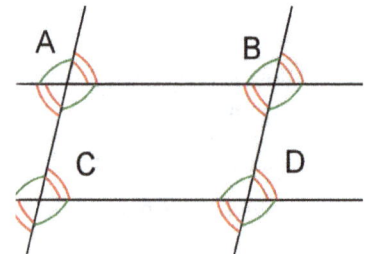

7. To solve this, we calculate how many whole times 8 inches goes into the 4 foot dimension (the one way) and how many whole times 5 3/4 inches goes into the 4 foot dimension (the other way). We care only about whole times because we're not dealing with fractions of puzzles. Those numbers tell us how many columns and rows of puzzles fit on the table. Lastly we simply multiply those numbers.

   Since 4 ft/8 in = 48 in/8 in = 6, exactly six columns of puzzles fit on the table.

   And since 4 ft/(5 ¾ in) = 48 in/(5 ¾ in) = 48/(23/4) = 48 · 4/23 ≈ 8.35, eight rows of puzzles fit on the table.

   In total, 6 · 8 = 48 puzzles fit on the table, in six columns and eight rows.

8. Check the student's drawing. When redrawn at a scale of 1:5, the dimensions of the picture will be 8/5 times larger than the dimensions in the drawing drawn at a scale of 1:8.

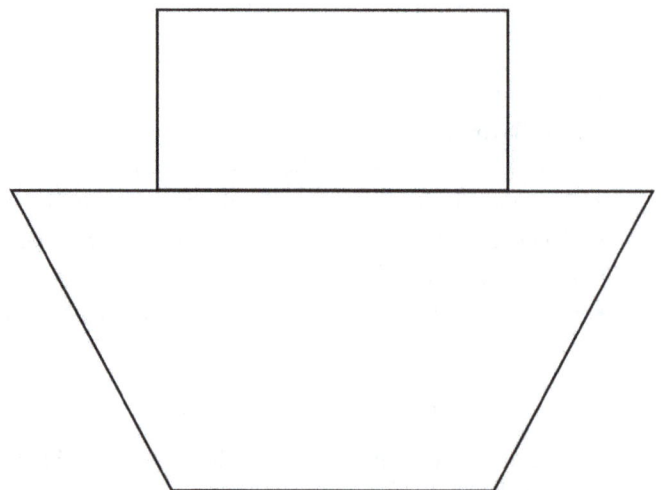

9. a. Check the student's triangle. It should be congruent to the one at the right (which is not to true scale). This means that the angles should measure the same, and the sides should be the same length as the triangle at the right, but possibly with a different orientation. Use a compass and straightedge and the method for drawing a triangle with three given sides explained in the lesson *Basic Geometric Constructions*.

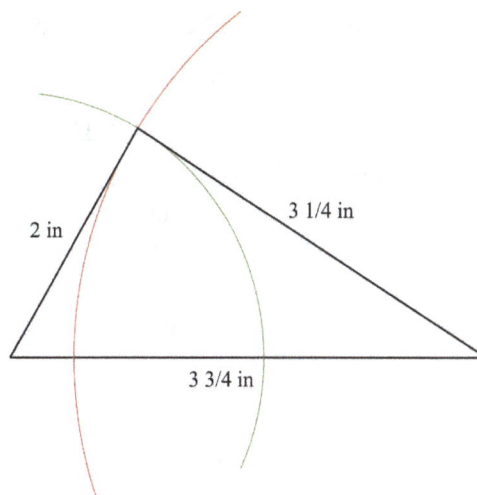

2 in · 3 1/4 in · 3 3/4 in

b. It is not possible. The sum of any two sides should be greater than the third side in order to create a triangle, but in this case, $3 + 3.5 < 7$. So, the lengths 7 cm, 3 cm, and 3.5 cm do not form a triangle.

10. a. The outcomes are listed on the right.

b. P(blue; blue) = 1/16

c. P(green; not green) = 3/16

d. P(not blue; yellow) = 3/16

e. P(yellow or green; red or blue) = 4/16 = 1/4

Sample space:

YY, YR, YB, YG

RY, RR, RB, RG

BY, BR, BB, BG

GY, GR, GB, GG

11. a. Answers will vary. Check the student's answer. For example: The coins land on the moon.

b. Answers will vary. Correct answers include:

- You get one heads and one tails (in either order).

- Both coins show different faces.

- Both coins show the same face.

12. No, John's conclusion is not correct. It is perfectly normal for the frequencies to vary when a chance experiment is repeated, and the frequencies 178 and 153 are definitely within normal variation when a die is rolled 1,000 times.

13. Sam's sampling method uses self-selection: the people choose whether they want to take part or not. For a sampling method to be random, it must use external selection where the respondents are chosen by the researcher. Moreover, Sam has no reason to believe that the people who happen to pass by the corner near his home are a representative sample of the people in the city as a whole.

Mixed Review 20, p. 131

1. In 2010, the factory sold about 850,000 candles and in 2015, about 1,050,000 candles. The percentage of increase is $(1,050,000 - 850,000)/850,000 \approx \underline{23.5\%}$.

2. The sides of the pentagons measure 1.9 cm and 1.5 cm, so the scale ratio is $1.5/1.9 = \underline{15:19}$. The scale factor is $15/19 \approx \underline{0.79}$.

3. a. No, they are not equal. The first rate, 60 words per 90 seconds, equals 60 words in 1 ½ minutes = 120 words in 3 minutes, whereas the second rate is 135 words in 3 minutes.

b. The unit rates are: 60 words/(1 ½ min) = 40 words/min and 135 words/3 min = 45 words/min. In five minutes, you will type 5 min · (45 words/min) − 5 min · (40 words/min) = 225 words − 200 words = $\underline{25 \text{ words more}}$, typing with the faster rate.

4. The tip is 0.05 · $40 = $2 and the sales tax is 0.068 · $40 = $2.72. The total cost is then $40 + $2 + $2.72 = <u>$44.72</u>.

5. a. The first is ($5 − $4.50)/$4.50 ≈ <u>11.1% more expensive</u> than the second.

   b. The second is ($5 − $4.50)/$5 = <u>10% cheaper</u> than the first.

6. A cube with 1-foot sides measures 1 cubic foot. For that, we need 1-foot or 12-inch edges. We would need six of the little cubes along the width, height, and depth to make the cube with 1-foot sizes, so that means we need 6^3 = <u>216 of the little cubes</u>.

   Another solution: 1 cubic foot is $(12 \text{ in})^3$ = 1,728 cubic inches. The small cube is $(2 \text{ in})^3$ = 8 cubic inches. So, we need 1,728/8 = <u>216 of the little cubes to make a cubic foot</u>.

7.

a.	$\begin{aligned} \dfrac{x}{5} &= -4.08 \\ x &= -4.08 \cdot 5 \\ x &= -20.4 \end{aligned}$	b.	$\begin{aligned} \dfrac{w}{-0.2} &= -0.4 \\ w &= -0.4 \cdot (-0.2) \\ w &= 0.08 \end{aligned}$
c.	$\begin{aligned} 2x + 7 &= -4(x + 5) \\ 2x + 7 &= -4x - 20 \\ 2x + 4x &= -20 - 7 \\ 6x &= -27 \\ x &= -27/6 = -4\,\tfrac{1}{2} \end{aligned}$	d.	$\begin{aligned} \dfrac{x + 1}{4} &= -2 \\ x + 1 &= -8 \\ x &= -9 \end{aligned}$

8. The principal's sampling method is biased, because not every parent has an equal chance of being selected into the sample. The principal surveyed only the parents that were attending a basketball game. Those parents may be more likely to want to improve the sports facilities than to support the other options, and therefore his sample is likely to be biased.

9. The dot plots show the amount of sugar in fruit and pop drinks (250 ml portion of drink).

   a. Based on the plots, both types of drinks tend to contain a similar amount of sugar.

   There are two ways to tell that from the plots:

   (1) The distributions overlap a lot: the fruit drinks contain 20-30 grams of sugar per 250 ml, and the pop drinks contain 18-32 grams.

   (2) We can easily figure out the medians using the plots, and they are very close: 25 grams for the fruit drinks and 26 grams for the pop drinks.

   b. The sugar amount in pop drinks varies somewhat more than the sugar amount in fruit drinks. We can tell that from the range: the pop drinks have a range of 14 grams and the fruit drinks have a range of 10 grams.

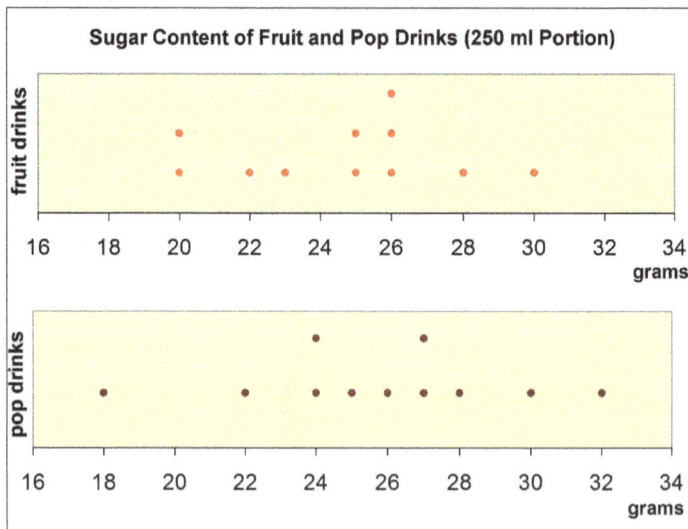

Sugar Content of Fruit and Pop Drinks (250 ml Portion)

10. Let x be the distance AB. Since ABC is (at least approximately) a right triangle, we can calculate x by using the Pythagorean Theorem:

$$2^2 + 2^2 = x^2$$
$$4 + 4 = x^2$$
$$x^2 = 8$$
$$x = \sqrt{8} \approx 2.83 \text{ mi} \quad \text{(We ignore the negative root.)}$$

Going from A to C and then from C to B is 4 miles. This is 4 mi − 2.83 mi = <u>1.17 miles longer</u> than going directly from A to B.

11. The sample space is below. It is of course possible to make a tree diagram, but that will be fairly large, so I chose to represent the sample space with a table in order to take less space.

The table represents the individual socks with labels from B1 through W6, the first letter indicating the color. A dash "-" indicates impossible outcomes. For example, if the first sock is B1, then the second sock cannot be the same sock, B1, so the cell for B1B1 shows a dash. There are 12 · 13 = 156 possible outcomes. The highlighted cells show the outcomes where both socks are the same color.

First → Second ↓	B1	B2	B3	B4	B5	R1	R2	W1	W2	W3	W4	W5	W6
B1	-	BB	BB	BB	BB	RB	RB	WB	WB	WB	WB	WB	WB
B2	BB	-	BB	BB	BB	RB	RB	WB	WB	WB	WB	WB	WB
B3	BB	BB	-	BB	BB	RB	RB	WB	WB	WB	WB	WB	WB
B4	BB	BB	BB	-	BB	RB	RB	WB	WB	WB	WB	WB	WB
B5	BB	BB	BB	BB	-	RB	RB	WB	WB	WB	WB	WB	WB
R1	BR	BR	BR	BR	BR	-	RR	WR	WR	WR	WR	WR	WR
R2	BR	BR	BR	BR	BR	RR	-	WR	WR	WR	WR	WR	WR
W1	BW	BW	BW	BW	BW	RW	RW	-	WW	WW	WW	WW	WW
W2	BW	BW	BW	BW	BW	RW	RW	WW	-	WW	WW	WW	WW
W3	BW	BW	BW	BW	BW	RW	RW	WW	WW	-	WW	WW	WW
W4	BW	BW	BW	BW	BW	RW	RW	WW	WW	WW	-	WW	WW
W5	BW	BW	BW	BW	BW	RW	RW	WW	WW	WW	WW	-	WW
W6	BW	BW	BW	BW	BW	RW	RW	WW	WW	WW	WW	WW	-

a. P(first sock white) = 6/13 ≈ 46.2%

b. P(WW) = 30/156 ≈ 19.2%

c. P(two socks of same color) = (20 + 2 + 30)/156 = 52/156 ≈ 33.3%

Please see the file for the End of the Year Test for grading instructions.

Integers

1. Answers will vary. Check the student's answer. $-15 + 10 = -5$. For example: A fish swimming at a depth of 15 ft rose 10 ft, and now it is 5 ft below the surface. Or, Mary owed her mom $15. She paid back $10 of her debt, and now she only owes her mom $5. Or, the temperature was $-15°$. It rose 10 degrees and now the temperature is $-5°$.

2. Answers will vary. Check the student's answer. $4 \cdot (-2) = -8$. For example: A certain ion has a charge of -2. Four such ions have a charge of -8. Or, four students bought ice cream for $2 each, but none of them had any money with them. Each of them borrowed $2 from a teacher. Now, their total debt is $8. Or, a stick reaches 2 m below the surface of the lake. If we put four such sticks end-to-end, they will reach to the depth of 8 m below the surface.

3. a.

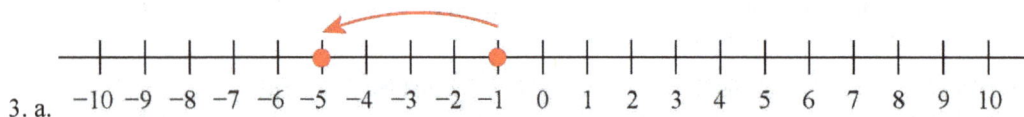

b.

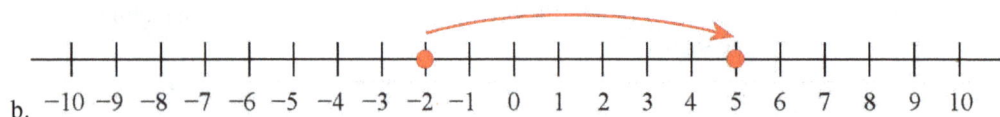

c.

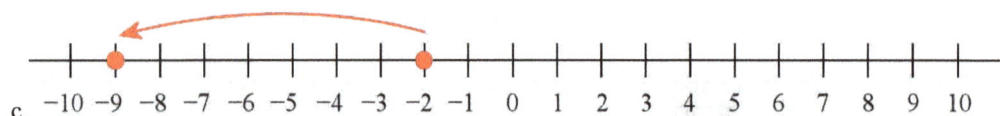

4. a. 2   b. -1 c. 25   d. 24   e. -12 f. 12

5. $|-5 - (-15)| = |10| = 10$.

6. a. $-1/8$ b. $-1/4$ c. 4 1/5

Rational Numbers

7. a. 1 1/28   b. $45.8\overline{3}$
   c. 0.00077   d. 0.0144
   e. 1 4/5   f. -6 2/7
   g. -0.2 or $-1/5$ h. 4

See below full solutions for 7. **g.** and 7. **h.** since they involve both a fraction and a decimal.

g. $-\dfrac{1}{6} \cdot 1.2$ If we use fraction arithmetic, this becomes: $= -\dfrac{1}{6} \cdot \dfrac{12}{10}$ $= -\dfrac{1}{6} \cdot \dfrac{6}{5} = -\dfrac{1}{5}$ If we use decimal arithmetic, we get $-\dfrac{1}{6} \cdot 1.2 = 1.2 \div (-6) = -0.2$	h. $-\dfrac{2}{5} \div (-0.1)$ If we use decimal arithmetic, this becomes $-0.4 \div 0.1 = 4$ (because $4 \cdot 0.1 = 0.4$). With fraction arithmetic, we get $-\dfrac{2}{5} \div \left(-\dfrac{1}{10}\right)$ $= \dfrac{2}{5} \cdot \dfrac{10}{1} = 4$

8. a. 1748/10,000   b. −483/100,000   c. 2 43928/1,000,000

9. a. −0.0028   b. 24.93   c. 7.01338

10. a. 0.53846   b. 1.$\overline{81}$

11.

a. $1.2 \cdot 25 = 30$ Answers will vary. Check the student's answer. For example: The price of a pair of scissors costing \$25 is increased by 20%. The new price is \$30. Or, a line segment that is 25 cm long is scaled by a scale factor 1.2, and it becomes 30 cm long. Or, the lunch break, which used to be 25 minutes long, is increased by 1/5. Now it is 30 minutes long.
b. $(3/5) \div 4 = (3/5) \cdot (1/4) = 3/20$. Answers will vary. Check the student's answer. For example: There is 3/5 of a large pizza left, and four people share it equally. Each person gets 3/20 of the original pizza. Or, a plot of land that is 3/5 square mile is divided evenly into four parts. Each of the parts is 3/20 square mile $= 15/100$ sq. mi. $= 0.15$ sq. mi.

Algebra

12.

a. $15s - 10$	b. $5x^4$	c. $3a + 3b - 6$
d. $1.02x$	e. $2w - 4$	f. $-3.9a + 0.5$

13.

a. $7x + 14$ $= 7(x + 2)$	b. $15 - 5y$ $= 5(3 - y)$	c. $21a + 24b - 9$ $= 3(7a + 8b - 3)$

14.

a. $\begin{aligned} 2x - 7 &= -6 \\ 2x &= 1 \\ x &= 1/2 \end{aligned}$	b. $\begin{aligned} 2 - 9 &= -z + 4 \\ -7 &= -z + 4 \\ -11 &= -z \\ z &= 11 \end{aligned}$
c. $\begin{aligned} 120 &= \frac{c}{-10} \\ -1200 &= c \\ c &= -1200 \end{aligned}$	d. $\begin{aligned} 2(x + \tfrac{1}{2}) &= -15 \\ 2x + 1 &= -15 \\ 2x &= -16 \\ x &= -8 \end{aligned}$
e. $\begin{aligned} \frac{2}{3}x &= 266 \\ 2x &= 798 \\ x &= 399 \end{aligned}$	f. $\begin{aligned} x + 1\frac{1}{2} &= \frac{3}{8} \\ x &= \frac{3}{8} - 1\frac{1}{2} \\ x &= \frac{3}{8} - \frac{12}{8} = -\frac{9}{8} = -1\frac{1}{8} \end{aligned}$

15. From the formula $d = vt$ we can find that $t = d/v$. In this case, $t = 0.8$ km/(12 km/h) $= 0.0\overline{6}$ h $= 0.0\overline{6}$ h $\cdot$ (60 min/h) $= \underline{4\ minutes}$. This is reasonable because the distance he ran is fairly short.

16. a. The equation that matches the situation is $\dfrac{4p}{5} = 48$.

   b. $\dfrac{4p}{5} = 48$

   $4p = 240$

   $p = 60$

   The original price was $60.

17. Let w be the width of the rectangle. The student can write any of the equations below:

   - $2w + 2 \cdot 55 = 254$
   - $2w + 110 = 254$
   - $w + w + 55 + 55 = 254$
   - $w + w + 110 = 254$
   - $w + 55 + w + 55 = 254$

   A solution of the equation:

   $$2w + 110 = 254$$
   $$2w = 144$$
   $$w = 72$$

   The rectangle is $\underline{72\ cm}$ wide.

18.

   a.
   $$3x - 7 < 83$$
   $$3x < 90$$
   $$x < 30$$

   b.
   $$2x - 16.3 \geq 10.5$$
   $$2x \geq 26.8$$
   $$x \geq 13.4$$

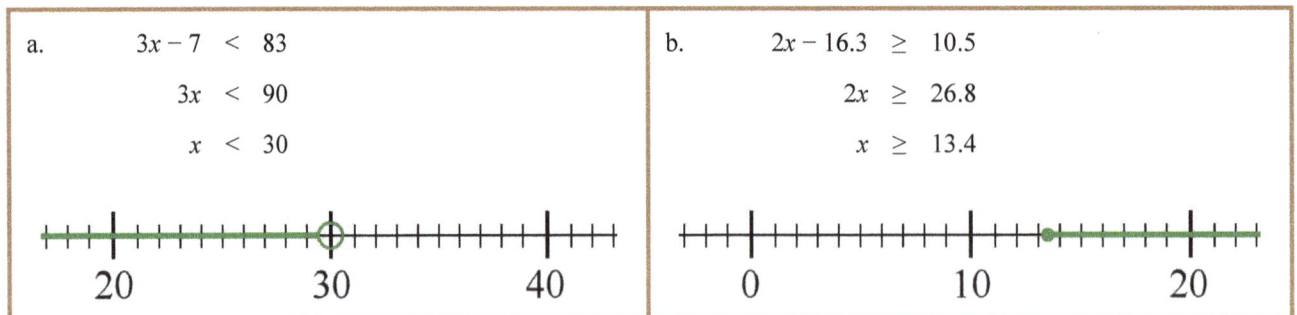

19. a. Let n be the number of boxes. The cost of the boxes with the discount is $15n - 25$.
   The inequality is $15n - 25 \leq 150$. Solution:

   $$15n - 25 \leq 150$$
   $$15n \leq 175$$
   $$n \leq 11.67$$

   b. The solution means that you can buy 11 boxes at most.

20.

a.	b.
$9y - 2 + y = 5y + 10$ $10y = 5y + 10 + 2$ $5y = 12$ $y = 12/5 = 2\ 2/5$	$2(x + 7) = 3(x - 6)$ $2x + 14 = 3x - 18$ $2x - 3x = -18 - 14$ $-x = -32$ $x = 32$
c. $\dfrac{y + 6}{-2} = -10$ $y + 6 = 20$ $y = 14$	d. $\dfrac{w}{2} - 3 = 0.8$ $\dfrac{w}{2} = 3.8$ $w = 7.6$

21. See the graph on the right.

22. a. See the graph on the right.

    b. The slope is −2.

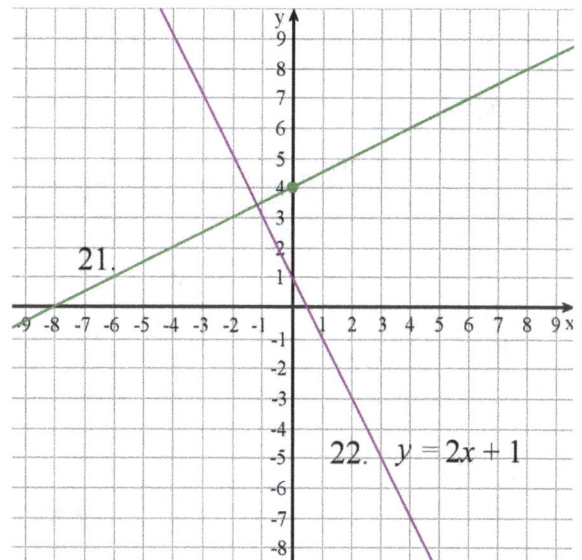

21.

22. $y = 2x + 1$

Ratios, Proportions, and Percent

23.

a. Lily paid $6 for 3/8 lb of nuts.

$$\frac{\$6}{\frac{3}{8}\ \text{lb}} = \$6 \cdot \frac{8}{3}\ \text{per lb} = \$16\ \text{per lb}$$

b. Ryan walked 2 ½ miles in 3/4 of an hour.

$$\frac{2\frac{1}{2}\ \text{mi}}{\frac{3}{4}\ \text{h}} = \frac{5}{2} \cdot \frac{4}{3}\ \text{mi/h} = \frac{20}{6}\ \text{mi/h} = 3\frac{1}{3}\ \text{mi/h}$$

24. The graph below shows the distance covered by a moped advancing at a constant speed.

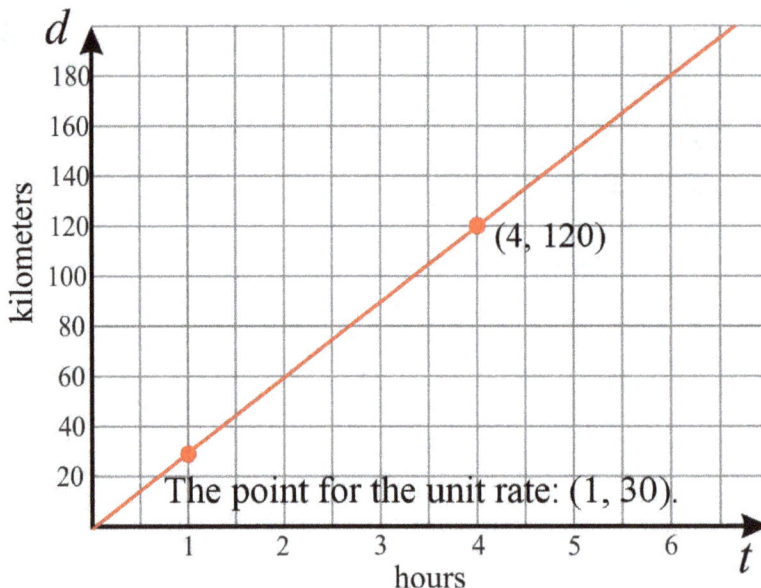

The point for the unit rate: (1, 30).

   a. The speed of the moped is 30 km/h.

   b. See the image above. The point is (4, 120), and it signifies that after driving 4 hours, the moped has covered 120 km.

   c. $d = 30t$

   d. See the image. It is the point (1, 30).

25. a. The Toyota Prius gets better gas mileage, because it gets 565 mi/11.9 gal ≈ 47.48 mi/gal whereas a Honda Accord gets 619 mi/17.2 gal ≈ 35.99 mi/gal.

   b. The cost of driving 300 miles with a Toyota Prius is 300 mi · (11.9 gal/565 mi) · $3.19/gal ≈ $20.16.
The cost of driving 300 miles with a Honda Accord is 300 mi · (17.2 gal/619 mi) · $3.19/gal ≈ $26.59.
The difference is $26.59 − $20.16 = $6.43.

26. She can withdraw $2,500 · 0.08 · 3 + $2,500 = $600 + $2,500 = $3,100.

27. After the 15% price increase, the ticket costs 1.15 · $10 = $11.50.
Then, the price decreased by 25% is 0.75 · $11.50 = $8.625 ≈ $8.63.

28. a. The percentage of increase was (72,000 − 51,500)/51,500 ≈ 39.8%.

   b. She will have 1.398 · 72,000 = 100,656 visitors ≈ 101,000 visitors.

29. Let r be the amount of rainfall in the previous month. Then, $1.35r = 10.5$ cm, from which $r = 10.5$ cm/1.35 ≈ 7.8 cm.

30. The side of the enlarged square is (4/3) · 15 cm = 20 cm. Its area is 20 cm · 20 cm = 400 cm^2.

31. There are two basic ways to calculate the distance on the map from the distance in reality. One way is that we first convert the given distance, 8 km, into centimeters, which are units used on the map, and then multiply by the ratio 1:50,000.

8 km = 8,000 m = 800,000 cm. The distance on the map is 800,000 cm · (1/50,000) = 16 cm.

Another way is to convert the ratio so that it uses common measuring units. The ratio 1:50,000 signifies that 1 cm on the map is 50,000 cm in reality. From this, we can write 1 cm = 50,000 cm = 500 m = 0.5 km. So 1 cm on the map corresponds to 0.5 km in reality. The given distance of 8 km corresponds to 8 km ÷ (0.5 km/1 cm) = 16 cm.

32. Proportions vary as there are several different ways to write the proportion correctly. Here are four of the correct ways. Besides these four, you will get four more by switching the right and left sides of these four equations.

$\dfrac{600 \text{ ml}}{554 \text{ g}} = \dfrac{5000 \text{ ml}}{x}$	$\dfrac{554 \text{ g}}{600 \text{ ml}} = \dfrac{x}{5000 \text{ ml}}$	$\dfrac{5000 \text{ ml}}{600 \text{ ml}} = \dfrac{x}{554 \text{ g}}$	$\dfrac{600 \text{ ml}}{5000 \text{ ml}} = \dfrac{554 \text{ g}}{x}$

The key point is that in each of the correct ways, x ends up being multiplied by 600 ml in the cross-multiplication. If x ends up being multiplied by 554 g or 5,000 ml in the cross-multiplication, the proportion is set up incorrectly.

Here is the solution process for one of the proportions above. Each of the others has the same final solution, $\underline{x = 4{,}617 \text{ g}}$.

$$\frac{600 \text{ ml}}{554 \text{ g}} = \frac{5000 \text{ ml}}{x}$$

$$600 \text{ ml} \cdot x \qquad 554 \text{ g} \cdot 5000 \text{ ml}$$

$$x = \frac{554 \text{ g} \cdot 5000 \text{ ml}}{600 \text{ ml}}$$

$$x = 4{,}617 \text{ g}$$

33. A farmer sells potatoes in sacks of various weights. The table shows the price per weight.

Weight	5 lb	10 lb	15 lb	20 lb	30 lb	50 lb
Price	$4	$7.50	$9	$12	$15	$25

a. These two quantities are *not* in proportion. For example, looking at the cost of potatoes for 5 lb and for 20 lb, the weight increases four-fold, but the cost increases only three-fold (from $4 to $12). Or, when the weight increases three-fold from 5 lb to 15 lb, the price does not increase three-fold but, from $4 to $9.

Another way to see that is in the beginning of the chart, the weights increase by 5 lb up to 20 lb, but the cost does not increase by the same amount. Instead, the cost increases first by $3.50, then by $1.50, then by $3.

b. There is no need to answer this, since the quantities are not in proportion.

Geometry

34. a. The given rectangle measures 7 cm by 10 cm. We multiply those by 45 to get the true dimensions:
7 cm · 45 = 315 cm = 3.15 m and 10 cm · 45 = 450 cm = 4.5 m.
The area is A = 3.15 m · 4.5 m = 14.175 m^2.

    b. The dimensions of the room at a scale 1:60 will be 45/60 = 3/4 of the dimensions of the room drawn at a scale of 1:45 so the scale drawing will be smaller than the drawing given in the problem. The width is (3/4) · 7 cm = 5.25 cm and the height is (3/4) · 10 cm = 7.5 cm.

    Or, you can divide the actual dimensions by 60 to get 315 cm ÷ 60 = 5.25 cm and 450 cm ÷ 60 = 7.5 cm.

Scale 1:45

Scale 1:60

35. We simply multiply the given dimensions (which are in inches) by the ratio 3 ft/1 in, essentially multiplying them by 3:

    4 ¼ in · (3 ft/1 in) = 12 3/4 ft or 12 ft 9 in and

    3 ½ in · (3 ft/1 in) = 10 ½ ft or 10 ft 6 in.

36. A = π · (8 cm)^2 ≈ 201 cm^2.

37. C = π · 2 · 9 in = 56.5 in.

38. Check the student's drawing. The image below is not to true scale, but the student's drawing of a triangle should have the same shape as the triangle below. It will just be larger.

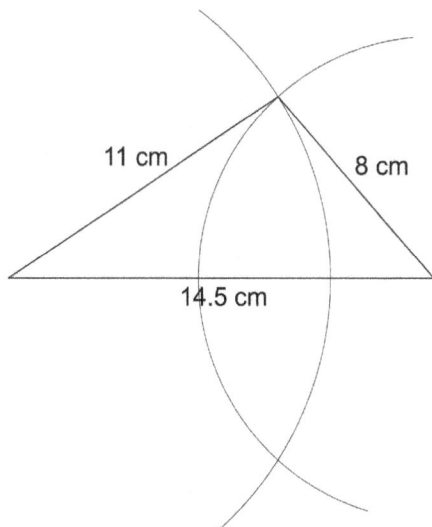

39. a. No, it doesn't.

b. The three given angles determine the shape of the triangle. The 8-cm side can be opposite of any of the given angles, so we get three different triangles. The images below are not to true scale but are smaller than in reality. They give you an idea of what the three different triangles look like. Check the student's drawings.

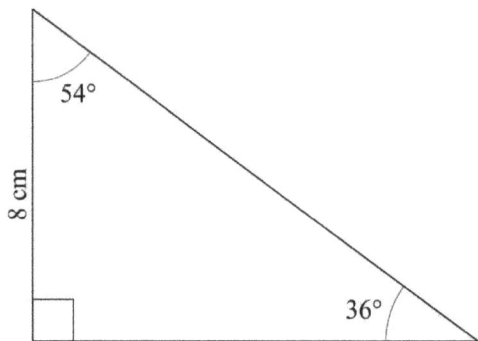

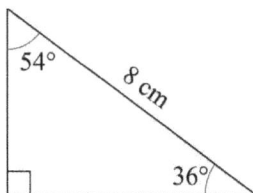

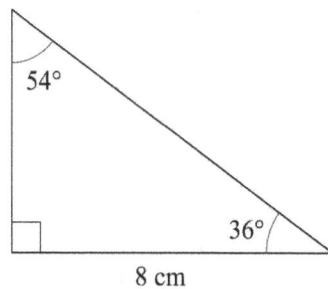

40. a. There are several ways to write an equation for x.

(1) Since angle x and the 53° angle are supplementary, $x + 53° = 180°$, from which $x = 180° − 53° = 127°$.

(2) Angle y and the 53° angle are vertical angles, so $y = 53°$. Then, angle x and angle y are supplementary, so we can write the equation $x + 53° = 180°$, and once again $x = 180° − 53° = 127°$.

b. There are several ways to write an equation for z.

(1) Since angles y, z, and the 18° angle lie along the same line (line l), their measures sum up to 180°, and we can write $y + z + 18° = 180°$. Since y and the 53° angle are vertical angles, $y = 53°$ and we can substitute that for y in the equation to get:

$$53° + z + 18° = 180°$$
$$z = 180° − 53° − 18°$$
$$z = 109°$$

(2) Or, since the combination angle $z + 18°$ and x are vertical angles, $z + 18° = x$. From part (a) we know that $x = 127°$ so the equation becomes:

$$z + 18° = 127°$$
$$z = 127° − 18°$$
$$z = 109°$$

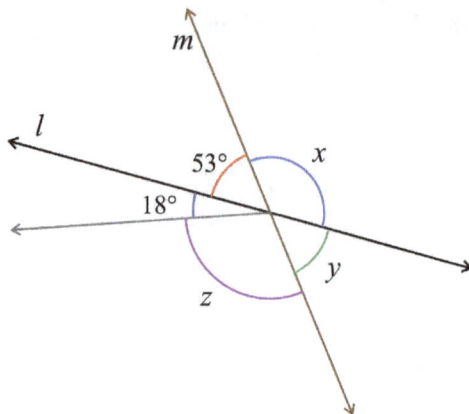

41. The triangle in the image has an angle of 43° and a right angle, so its third angle is $180° − 43° − 90° = 47°$.

The unknown angle x supplements the third angle of the triangle, so its measure is $180° − 47° = \underline{133°}$.

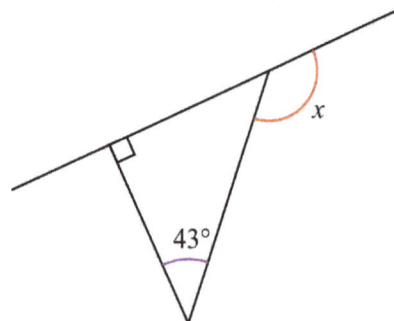

42. a. The cross section is a rectangle.

b. The cross section is a triangle .

c. The cross section is a trapezoid.

43. a. The roof or top part of the entire canopy is a triangular prism. The base of the prism is a triangle with a base side of 10 ft and a height of 4 ft, so its area is 10 ft · 4 ft / 2 = 20 ft^2.

The volume is $V = A_b \cdot h = 20$ ft^2 · 14 ft = 280 ft^3.

b. The bottom part is a rectangular prism, and its volume is 10 ft · 14 ft · 8 ft = 1,120 ft^3.

The total volume is 280 ft^3 + 1,120 ft^3 = $\underline{1,400\text{ ft}^3}$.

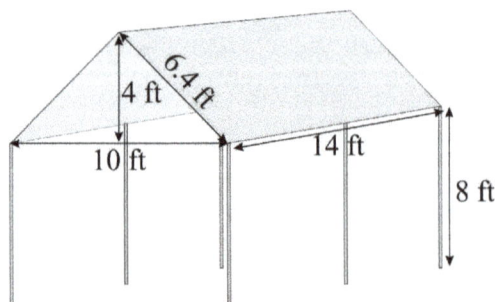

44. a. One way is to use the formula for the area of a trapezoid. See the image on the right.
The area of one trapezoid is A = (10 cm + 15 cm)/2 · 7.5 cm = 93.75 cm².

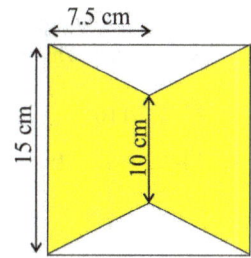

Then, the area of the two trapezoids is 2 · 93.75 cm² = 187.5 cm².

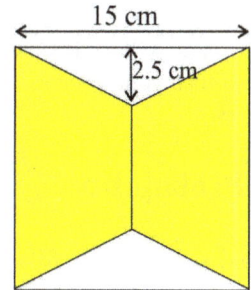

Another way is to subtract the area of the two white triangles from the area
of the entire 15 cm by 15 cm square.

The area of one triangle is 15 cm · 2.5 cm/2 = 18.75 cm².

The area of the two trapezoids is then 15 cm · 15 cm − 2 · 18.75 cm² = 187.5 cm².

b. The trapezoids cover 187.5/(15 · 15) ≈ **83.3%** of the entire square.

45. a. V = $A_b · h$ = π · (6 cm)² · 4.5 cm = 508.938 cm³ ≈ **509 cm³**.

b. The volume is 509 ml = 0.509 L.

46. a. Since 1 ft = 12 in, 1 cubic foot has 12 in · 12 in · 12 in = **1,728 in³**.

b. 3 ¼ ft = 3 · 12 in + 3 in = 39 in. The volume is V = (39 in)³ = **59,319 in³**.

The Pythagorean Theorem

47. a. Its area is 5 m², because (√5 m)(√5 m) = (√5 m)² = 5 m².

b. It is √45 cm ≈ 6.7 cm.

48. $57^2 + 76^2 \overset{?}{=} 95^2$

    $3{,}249 + 5{,}776 \overset{?}{=} 9{,}025$

             $9{,}025 = 9{,}025$

Yes, the lengths 57 cm, 95 cm, and 76 cm do form a right triangle.

49. Using the Pythagorean Theorem, we can write the equation

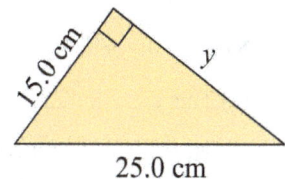

$$15^2 + y^2 = 25^2$$
$$225 + y^2 = 625$$
$$y^2 = 400$$
$$y = \sqrt{400} = 20 \text{ (We ignore the negative root.)}$$

The side is 20 cm long.

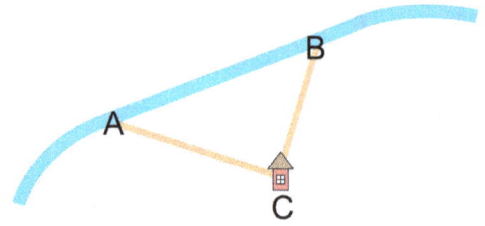

50. Let x be the length AB. Since ABC is a right triangle, applying the Pythagorean Theorem we get:

$$120^2 + 110^2 = x^2$$
$$14400 + 12100 = x^2$$
$$x^2 = 26{,}500$$
$$y = \sqrt{26{,}500} = 162.788 \text{ m} \quad \text{(We ignore the negative root.)}$$

You will walk 120 m + 110 m = 230 m. Your friends walk 163 m. So, you will walk 230 m − 163 = <u>67 m</u> more than your friends.

Probability

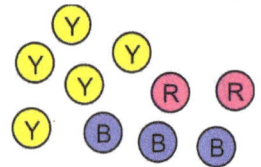

51. a. P(not red) = 8/10 = 4/5

   b. P(blue or red) = 5/10 = 1/2

   c. P(green) = 0

52. a.

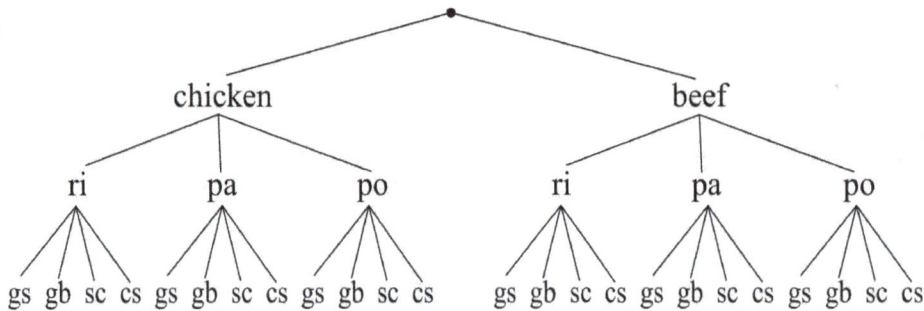

   b. P(beef, rice, coleslaw) = 1/24

   c. P(no coleslaw nor steamed cabbage) = 12/24 = 1/2

   d. P(chicken, green salad) = 3/24 = 1/8

53. None of the conclusions (a), (b), or (c) are valid.

   (a) This die is unfair.

   Not valid. In a repeated random experiment, the frequencies for the various outcomes do vary, and the variability seen in the chart is definitely within normal variation. In fact, the data comes from running a computer simulation that uses random numbers, and the simulation was run 1,000 times.

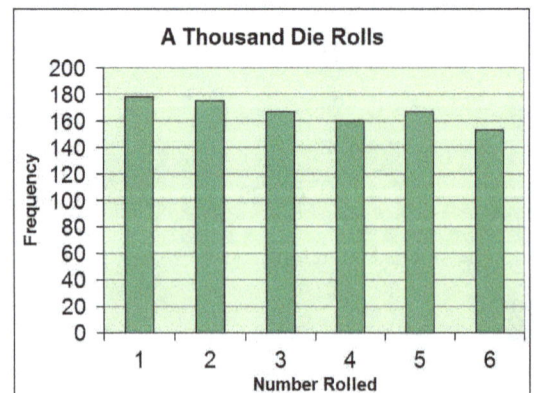

   (b) On this die, you will always get more 1s than 6s.

   Not valid. The die is not necessarily unfair. A normal die could produce the frequencies seen in the chart.

   (c) Next time you roll, you will not get a 4.

   Not valid. Rolling a die is a random experiment and you might get 4 the next time you roll.

54. We can toss 10 coins (or a single coin 10 times) to simulate 10 children being born. Let heads = girl, and tails = boy (or vice versa).

    Then, repeat that experiment (tossing 10 coins) hundreds of times. Observe how many of those repetitions include 9 heads and one tail , which means getting 9 girls and one boy. The relative frequency is the number of times you got 9 heads and one tail divided by the number of repetitions, and gives you an approximate value for the probability of 9 girls and 1 boy in 10 births.

Statistics

55. Cindy's sampling method is <u>biased</u>. She chose students from her class, which means that all the other students in the college didn't have a chance to be selected in her sample. For a sampling method to be unbiased, every member of the population has to have an equal chance of being selected in the sample. (Her method would work if she was only studying the students in her class.)

56. Four people are running for mayor in a town of about 20,000 people. Three polls were conducted, each time asking 150 people who they would vote for. The table shows the results.

	Clark	Taylor	Thomas	Wright	Totals
Poll 1	58	19	61	12	150
Poll 2	68	17	56	9	150
Poll 3	65	22	53	10	150

    a. Based on the polls, we can predict <u>Clark</u> to be the winner of the election. He is leading in two of the three polls.
    b. To estimate how many votes Thomas will get, use the average percentage of votes he got in the three polls. You can calculate that as $((61 + 56 + 53) \div 3)/150$ or as $(61/150 + 56/150 + 53/150) \div 3$. Either way, you will get 37.78%. This gives us the estimate that he would get $0.3778 \cdot 8{,}500 \approx$ <u>3,200 votes</u> in the actual election.
    c. We will gauge how much off the estimate of 3,200 votes is by using the individual poll results.
    Based on poll 1, we would estimate Thomas to get $(61/150) \cdot 8{,}500 \approx 3{,}460$ votes.
    Based on poll 2, we would estimate Thomas to get $(56/150) \cdot 8{,}500 \approx 3{,}170$ votes.
    Based on poll 3, we would estimate Thomas to get $(53/150) \cdot 8{,}500 \approx 3{,}000$ votes.

    Looking at the highest and lowest numbers (3,000 and 3,460), we can gauge that our estimate of 3,200 votes might be <u>off by a few hundred votes</u>.

57. The total number of people Gabriel surveyed is $45 + 57 + 18 = 120$. Of those, $45/120 = 37.5\%$ support building the highway.

    This gives us the estimate that $0.375 \cdot 2{,}120 = 795 \approx$ <u>800 households</u> in the community would support building the highway.

Opinion	Number
Support the highway	45
Do not support it	57
No opinion	18

58. a. Group 1 appears to have lost more weight.
    b. Group 1 appears to have a greater variability in the amount of weight lost.
    c. The person gained 1 pound.
    d. Yes, the method used with group 1 is significantly better than the other.

    The median weight loss for Group 1 is 10 lb whereas for Group 2 only a little over 4 lb. The difference in the medians is about 6 lb. The interquartile ranges are about 5 lb for both Groups. The difference in the medians (about 6 lb) is more than one time the interquartile range (about 5 lb), which shows us that the difference is significant.

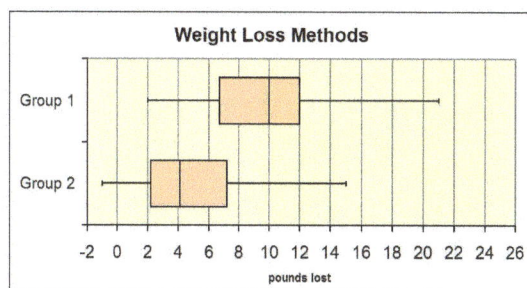

    Weight Loss Methods

    Group 1

    Group 2

    -2 0 2 4 6 8 10 12 14 16 18 20 22 24 26
    pounds lost

More from math MAMMOTH

Math Mammoth has a variety of resources to fit your needs. All are available as economical downloads, and most also as printed copies.

- **Math Mammoth Light Blue Series**
  A complete curriculum for grades 1-7. Each grade level includes two student worktexts (A and B), which contain all the instruction and exercises all in the same book, answer keys, tests, cumulative reviews, and a worksheet maker. International (all metric), Canadian, and South African versions are also available.

  https://www.MathMammoth.com/complete-curriculum

  https://www.MathMammoth.com/international/international

  https://www.MathMammoth.com/canada/

  https://www.MathMammoth.com/south_africa/

- **Math Mammoth Skills Review Workbooks**
  These workbooks are intended to be used alongside the Light Blue series full curriculum, and they provide additional review to the topics studied in the main curriculum, in a spiral manner.

  https://www.MathMammoth.com/skills_review_workbooks/

- **Math Mammoth Blue Series**
  Blue Series books are topical worktexts for grades 1-8, containing both instruction and exercises. They cover all elementary math topics from 1st through 7th grade and some for 8th grade. These books are not tied to grade levels, and are thus great for filling in gaps.

  https://www.MathMammoth.com/blue-series

- **Make It Real Learning**
  These activity workbooks concentrate on answering the question, "Where is math used in real life?" The series includes various workbooks for grades 3-12.

  https://www.MathMammoth.com/worksheets/mirl/

- **Review Workbooks**
  Workbooks for grades 1-7 that provide a comprehensive review of one grade level of math—for example, for review during school break or summer vacation.

  https://www.MathMammoth.com/review_workbooks/

Free gift!

- Receive over 350 free sample pages and worksheets from my books, plus other freebies:
  https://www.MathMammoth.com/worksheets/free

Lastly...

- Inspire4 is an inspirational website for the whole family I've been privileged to help with:
  https://www.inspire4.com